COUNT US IN

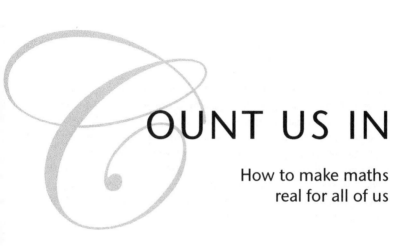

OUNT US IN

How to make maths
real for all of us

Gareth Ffowc Roberts

UNIVERSITY OF WALES PRESS
2016

www.uwp.co.uk

British Library Cataloguing-in-Publication Data.
A catalogue record for this book is available from the British Library.

ISBN   978-1-78316-796-8
eISBN   978-1-78316-797-5

The publication of this book, first published in Welsh by Gomer Press, Llandysul, Wales, is supported by the Welsh Books Council (WBC) and the Welsh Association of Lecturers in Mathematics in the Area Training Organisation (WALMATO).

Designed and typeset by Chris Bell, cbdesign
Printed by CPI Antony Rowe, Chippenham, Wiltshire
Front cover artwork by Valériane Leblond

To my grandchildren
Elis, Gwydion, Mari, Miriam, Olwen,
and their contemporaries

My wish is that you should learn in your native language.
So that no foreign Nation should have cause to mock you
on account of your Ignorance.

John Roberts, *Arithmetic: mewn Trefn Hawdd ac Eglur* (1768)

CONTENTS

Figures and plates		ix
Acknowledgements		xi
Preface		xiii
1	More cabbage, anyone?	1
2	Meeting of minds	7
3	'Nothing will come of nothing'	15
4	Setting the Recorde straight	21
5	'Neither a borrower nor a lender be'	33
6	Amazing Mayans	41
7	What do you reckon?	47
8	Prairie power	59
9	Putting down digital roots	67
10	Areas of (mis)understanding	79
11	Cracking the code	89
12	Does mathematics have a gender?	97
13	How to make maths real for all of us	111
Appendix		116
Answers to puzzles		120
Notes on chapters		125
Further reading		139
Index		141

FIGURES AND PLATES

Figures

Fig. 1. Mural of Einstein and Ramanujan at the Loyola
University College in Chennai, India   17

Fig. 2. The bust of Robert Recorde in St Mary's Church,
Tenby   23

Fig. 3. Woodcut from the title page of *The Ground of Artes*
(1543)   27

Fig. 4. Robert Recorde introduces the equals sign in
The Whetstone of Witte (1557)   30

Fig. 5. Caradog Jones's work on Pythagoras' theorem, 1880   65

Fig. 6. Celtic knot pattern on a cross near Llandeilo   69

Fig. 7. Hymn board in a Welsh chapel   73

Fig. 8. Slate plaque in Llanfechell, Anglesey, celebrating
the work of William Jones   81

Fig. 9. William Jones's use of π to denote the ratio of the
diameter of a circle to its circumference   83

Fig. 10. Llinos's present to her parents   98

Fig. 11. Mary Wynne Warner (1932–98)   102

Plates

Plate 1. Gutun Owain's use of Hindu–Arabic numerals in 1488/9

Plate 2. *Magical Symbols*, an interpretation by Anne Gregson
showing the use of Recorde's sign of equality

Plate 3. Representation of a date using glyphs on a Mayan
calendar

Plate 4. Pupils of the Rawson school in Patagonia, 1880

Plate 5. Star pattern on a Welsh quilt, 1875

Plate 6. Portrait of William Jones (1674–1749) by William Hogarth (1740)

Plate 7. Tribute to William Jones by Jan Abas

Plate 8. *Equally Puzzled* by Claudia Williams

ACKNOWLEDGEMENTS

THE INSPIRATION for this book draws heavily on my experience as a mathematics adviser for the county of Gwynedd and I am indebted to very many school pupils, their teachers and my fellow workers in that county. I also benefited greatly from my period working with trainee teachers and my fellow lecturers at what was originally Bangor Normal College, now the School of Education at Bangor University. In writing this book, I have been strongly encouraged by many colleagues throughout Wales who have a passion for the teaching of mathematics. Their support has been an energising force.

I have received practical support, advice and inspiration from many individuals, including: Janet Abas, Colin Barker, Alex Bellos, Marian Davies, Luned González, Ceris Gruffudd, Llion Jones, Gwyn Lewis, Gwyn and Margaret Lloyd, Rhys Llwyd, Hywel Wyn Owen, Pablo Pappolla, Islwyn Parry, Morfudd Phillips, Nia Powell, Dafydd Price, Elizabeth and Gordon Roberts, Huw Alun Roberts, Ceri Subbe, Richard Talbot, Gareth and Margaret Tilsley, Gerald Warner, Bethan and Islwyn Williams, and Ellen and Ian Williams. I am also indebted to staff at the University of Wales Press for their encouragement and support.

My strongest supporters, as well as my sternest critics, have been my son Huw and my daughter Llinos, with the support of their partners, Bethan and Aled. Urged on by my wife Menna, they made valiant efforts to keep their father's feet firmly on the ground.

REFACE

I GREW UP IN Holywell, a market town in Flintshire located between an industrial Deeside belt and a rural hinterland. Our house stood close to the old A55 along which flowed a constant stream of cars during the busy summer months carrying holiday-makers on their way to the beaches of Rhyl, Colwyn Bay, Llandudno and further west. I spent many happy hours during those periods sitting at the side of the A55, writing down the car registration numbers. With what aim in mind, I know not. A 'sad case', but somehow there was a magnetic attraction in the idea of collecting numbers, just as, some years later, I went through a phase of collecting stamps and today's children collect Sylvanian toys or Panini stickers.

When it rained I would wander indoors to play with building kits: metal Meccano strips in particular, and, long before the days of Lego, small rubber bricks that fitted into each other. Did these early experiences with numbers and shapes spark off my interest in mathematics, or was there something in the genes even before then?

It took me a long time to realise that not everyone shared my interest in mathematics and even longer to realise that it wasn't generally recognised as an activity that took its place naturally and comfortably within Welsh culture. When people talked about the *pethe* (an umbrella term used in Welsh to encompass the notion of shared cultural values), they were always referring to literature, poetry, religion, music, art or dance. What about mathematics and the sciences? Didn't I have as much interest in the *pethe* as any-one else? Why, then, were others apparently unwilling to include mathematics and the sciences as part of their *pethe*?

This book is aimed, in part at least, at such people: those who are comfortable with the traditional view of culture, and Welsh culture in particular, and are slightly taken aback at the suggestion that it could also include mathematics, of all things. The ideal reader is someone who is completely opposed to the suggestion of allowing mathematics to trespass on his or her cultural comfort zone. The book will also be of interest to those who welcome this wider definition and wish to learn more about Welsh mathematical links.

The book takes the form of a collection of short stories based on my personal experience, linked together by a number of sub-themes. At the back of the book I have added notes to each chapter, including some further references. In an Appendix I have listed Welsh number words in both their traditional and modern forms. Each chapter includes a mathematical puzzle related to the chapter's content; answers and comments on the puzzles are included in the Appendix.

The book has grown from a talk that I gave in Welsh in 1992 at the National Eisteddfod, held in Aberystwyth that year. The title of that talk, loosely translated, was 'Who counts?' In 2012, twenty years later, I wrote a book, again in Welsh, that attempts to answer the original question, entitling it, again in loose translation, as 'We all count'. This current book aims to interpret the main ideas in the original for an English-reading audience.

The process of counting, like the process of communicating with words, is common to all societies worldwide but, just as there is a rich variety of languages, so also is there a rich variety in methods of counting and of recording numbers, methods that have developed over centuries to meet the needs of various groups of people. The needs of a community of hunters are not the same as those of a society that has developed agricultural skills or a society that depends on trading.

Our attitudes, as Welsh people, to number and to the language or languages of number are part of this rich cross-cultural mix. We need to understand aspects of the history of Welsh involvement with numbers in order to understand current attitudes. Yes, we all count, but not necessarily to the same end or under the same circumstances.

Gareth Ffowc Roberts, July 2015

MORE CABBAGE, ANYONE?

ARE we all guilty of having a certain streak of sadism? When I taught students who were being trained as primary-school teachers, I was often tempted to play an underhand trick on the new intake of freshers. At their first lecture I explained that it was important for me to know something about their current skills in mathematics. In order to do that, I intended to set a short maths test: 'Make sure that you've got a clean sheet of paper. I'll ask the questions slowly. Everybody ready? No cheating! First question ...' A deathly hush, everyone listening intently. And then, 'Write one word, just one, that expresses how you're feeling: what's going through your mind at this precise moment.' A sigh of relief as everyone gradually realised that my threat to set a test was just a cruel joke, and everyone was more than ready to summarise their feelings in one word.

It's no surprise that the overwhelming number of responses were negative. It was rare for anyone to write words such as 'fantastic' or 'confident'. On the contrary, words such as 'nervous', 'anxious' and 'uncertain' were used far more often. But the most common word by far – year after year – was 'panic' or 'PANIC!' I would have had a negative reaction had I threatened to set a test in any subject – science, language, history, geography, music and so on – but the reaction is more extreme, and the cries of anguish more heartfelt, in mathematics than in any other subject.

Why does mathematics provoke such an extreme reaction? Is it because the answers to questions in mathematics tend to be either right or wrong? Two threes (2 × 3) are 6, no arguing, no room for debate, the answer can't be 5 or 7. We are conditioned from a young age to think of mathematics as a subject in which there can be no discussion, no possibility of an alternative opinion. The subject has its own inbuilt authority that transfers to teachers who mark answers with a tick or a cross, or to parents who pass on their own uncertainty and worry to the next generation: 'You *still* can't remember seven eights (7 × 8)?'

This negativity isn't in our genes; we're not born hating mathematics. On the contrary, young children commonly express how much they like doing mathematics. I had been invited to give a talk to a literary society in south Wales and was staying overnight with my son and his family. Before leaving the house, Mari, my five-year-old granddaughter, asked me where I was going and why. 'Well', I answered, 'I'm going to the village hall to give a talk on maths.' 'Oh', said Mari, 'I like doing maths', and then, after a pause, 'And will you be dancing there?' Quite how she made an association between mathematics and dancing remains a mystery but Mari clearly regarded both activities as belonging to the set of things that she liked doing. The trick is to maintain that natural liking and to foster it with care.

For many people their first encounter with algebra was that point in their education when mathematics literally became too abstract and they were not assisted to come to terms with the change of gear. In the words of a tale that has done its rounds on social media: 'Dear algebra, please stop asking us to find your *x*. She's never coming back, and don't ask *y*.' For many, the *x* remained a perpetual mystery that was never demystified for us during the early years of our secondary education. This only served to reinforce the belief built up during our time at primary school that mathematics wasn't something to be understood; rather it was something to be done – and to be suffered.

The answer often given by children to the gentle parental probe, 'What did you do in school today?' is 'Oh, nothing!' In an effort to snooker her four-year-old daughter at the end of a morning at nursery school, one resourceful parent asked her, 'Now, tell me three things that you did at school this morning.' Not to be outmanoeuvred, the child replied, 'Nothing, nothing, nothing!' However, responses can occasionally be more forthcoming and revealing. This was what our then five-year-old daughter, Llinos, wrote down as her response to this evergreen question. Try to work out what she's done before reading further:

Llinos was having trouble writing some of her numerals correctly, reversing the 4 and 5 in this case – a common practice at this age. Some gentle probing on our part revealed that today's lesson had been 'doing add-ups' and the teacher had introduced 'carry one' as a new idea. Llinos had learned to write 'd' (*degau* – the Welsh for tens) and 'u' (*unedau* – the Welsh for units) at the top of the sum and had picked up that it was important to begin by adding the numbers under the 'u' before moving on to add the numbers under the 'd'. Because the lesson's aim had been to introduce 'carry one', Llinos had also used that in her example, although it didn't apply in this case. She finished off her work with a flourish by adding the tick, and, smiling broadly, turned to her parents for their approval. She had no understanding at all of what she'd been doing. She hadn't interpreted the number in the first line as being thirty-three (33) nor the second number as forty-five (45) – quite a challenge for a five-year-old – and didn't understand that her answer was eighty-eight (88). She was happy and content that she had accepted the authority of the teacher, her own tick providing the crowning seal of approval. 'That's very good!' was our only possible response.

'It's all very well for you' is what I'm often told. 'You didn't have any difficulty with maths at school. It wasn't like that for me.

I never understood what was going on.' Such sentiments are heart-felt and contain more than a grain of truth. Not that I've never had difficulty understanding mathematics but, somehow, I seem to have had sufficient confidence to plough on. When I must have been about six years old, I sat at a table in our back kitchen on a wet Saturday morning and set myself the task of writing out the numbers, beginning with 1, 2, 3 and so on. My aim was to write down *every* possible number! I can't remember how far I got – possibly somewhere in the hundreds – but I do remember the feeling as I gradually realised that I couldn't possibly finish the task and that numbers simply never come to an end. In that split second I got a fleeting glimpse of infinity, that something could go on and on 'for ever'.

At my infants school the sums, as far as I can remember them (there was no talk of 'mathematics' in those days), were pretty straightforward. I do, however, have a vivid memory of being distraught at seeing a big red cross alongside every sum in my copybook following one particular lesson. I must have misunderstood something pretty basic. It's significant that it's the memory of that particular morning that has remained.

In my next class, having moved from the Infants to Standard 1, the teaching was much more formal – we all sat in rows and, dipping our steel pens into the inkwells on our desks, we copied into our books whatever Miss Williams wrote in chalk on the blackboard. Here again, one particular lesson stands out: a lesson on 'long division', a pet hate for many. After going through one example on the board, Miss Williams told us to carry on by ourselves to do a dozen or so similar sums. I remember thinking, 'What's going on here? I haven't got a clue. What am I supposed to do?' I experienced my own moment of panic that morning, which wasn't helped by my noticing that the other children in the class were hard at work, apparently unconcerned by the challenge. This was my first experience of not understanding something in mathematics – Miss Williams's instructions had made no sense at all. The teacher expected us to be content with *knowing how*, no more than that, whereas I wanted to *understand why*. Up to that point I'd managed to *understand why* whenever we were given new sums to do – add-ups, take-aways and multiplications – but *understanding why* when faced with long division was completely beyond me. It soon became clear that that was the expectation: don't ask why, just get on with it and keep your head down.

Yesterday, as today, arithmetic includes the skills of addition (+), subtraction (–), multiplication (×) and division (÷). Yesterday, as today, arithmetic (and the more broadly based idea of numeracy) is vital to the development of full citizenship and includes the application of number in everyday life as well as across other parts of the school curriculum. 'Arithmetic is the inheritance of civilised nations' was the opinion expressed by the winner of an essay competition at the 1859 Wrexham National Eisteddfod, and that sentiment has been repeated consistently over the years.

If there is broad agreement regarding the importance of numeracy, there have been deep disagreements about the means to develop it. We all experience a constant tension between *knowing how* and *understanding why*. For some, *knowing how* is quite sufficient: the only aim of a maths lesson is to get to *know how* to get the answer. For others, *understanding why* is just as important, if not more so. 'Absolute nonsense!', answers the first group. 'Getting it right is the only thing that counts.' The two groups disagree fundamentally regarding the nature and purpose of maths. To which group do you belong?

Learning to repeat instructions parrot-fashion – rote learning – was how generations of children experienced sums. Is it at all surprising that so many people who were at primary school before, say, the 1960s have negative attitudes to the subject? Things improved greatly from the 1960s onwards with more emphasis on giving pupils practical experiences in the classroom and on encouraging the use of language in mathematics. But progress has been gradual and change from one generation to the next is necessarily slow.

As a mathematics adviser during the 1980s, I was invited by a primary-school head teacher to call in to talk with one of the teachers who was refusing point blank to adopt 'modern' methods.

I had an interesting conversation with the teacher, a man in his mid-fifties, who had been brought up on the traditional methods and saw no good reason to change: 'If it was good enough for me, it's good enough for today's children too.' As I probed further, with a certain amount of care and diplomacy, it became clear that the teacher himself did not understand the methods that he was passing on to his pupils – he *knew how* but did not *understand why.* I was dangerously close to undermining his professional self-respect; hadn't he been using these methods for decades without a single complaint? We did, however, manage to prise open some new windows in order to expand his perception but it's doubtful if I managed to convince him completely. Many of the children under his care are themselves parents by now and can see the very different experiences in mathematics that their children are enjoying compared with what was offered to them by this particular teacher.

Is it, therefore, any wonder that the attitudes displayed by adults towards mathematics tend to be polarised? A relatively small number are fascinated by the subject, delighting in its patterns and its insights. Others loathe it completely and are prepared to boast about their incompetence openly and publicly, often referring to unfortunate experiences at school – weekly mental tests, ineffectual teachers, nasty teachers. Is there any truth in the saying, 'Maths is like cabbage: you love it or you hate it, depending on how it was served up to you at school'?

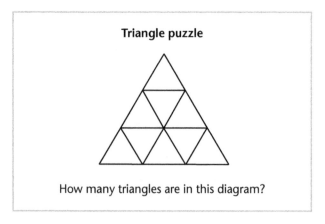

Triangle puzzle

How many triangles are in this diagram?

MEETING OF MINDS

QUÉBEC CITY bathes in sunlight as it welcomes an international conference to the green and lush campus of Laval University. In one of the modern lecture theatres delegates have gathered for a morning on the topic of 'ethnomathematics', and are welcomed at the outset by a small Maori choir greeting us in their native language. The choir's leader approaches the microphone to address the audience. '*Mes amis*', he begins, in a formal secondary-school French, deliberately showing his respect to the conference location and Québec state's main language. Only a few in the audience are fluent in French but we all understand his simple phrases and are moved by the sincerity of his message: 'The mountains of New Zealand greet the mountains of the state of Québec; our valleys greet your valleys; our rivers greet your rivers', pausing before his final greeting: 'And our *people* greet your *people*.' Another song from the choir and then, a good quarter of an hour into the meeting, one of the other members of the choir comes forward to present his paper.

By now the audience has been completely captivated by the simplicity and force of the presentation and is eager to hear more. This session is one of scores of others that form a conference of some 3,000 mathematics educators that are held once every four years in different cities across the world. One of the sub-themes of every conference is the link between mathematics and culture or rather the links between mathematics and the world's diverse cultures, as it becomes clear that mathematics is not a single body of unchallengeable knowledge. Rather, it is interpreted in differing ways by differing cultures, each through its own cultural prism. That day, it was the turn of the Maori to present their particular prism and to show how it influenced the teaching of mathematics in the schools of New Zealand.

There are (at least) two aspects to the relationship between mathematics and culture. On the one hand, the sparse 'beauty' of mathematics can spark within us a cultural response having the same quality as our response to, say, a Shakespearean sonnet, a sonata by Schubert or a sculpture by Michelangelo. The perfection of Pythagoras' theorem, for example, can satisfy us at our deepest levels of emotional understanding.

On the other hand, and at a different level, we use practical mathematical ideas on a daily basis – in our homes, with other people, on the high street and at work. We are taught these ideas at school but our interpretation of them is also influenced by our everyday experiences at home and in the community: they are mediated by our culture, and through the language or languages of that culture.

Mathematics is therefore not only a universal cultural phenomenon but also a product of our daily activities. In this latter respect local cultures and their languages are core factors as children grapple with basic mathematical ideas and as adults use those ideas in their everyday lives.

A specific example can help us reflect on the tensions between these two perspectives. One of the long-term concerns of the Maori has been that their children appear to have underachieved in their school mathematics by comparison with other New Zealanders, using the country's standard tests as a yardstick. For over a hundred and fifty years mathematics in New Zealand had been taught only through the medium of English, using curricula and textbooks that were rooted in the majority non-Maori culture, and taught in schools that didn't recognise the values basic to the Maori culture. This led inevitably to a situation where the Maori viewed many school subjects – and mathematics in particular – as being foreign to their natural culture. The phrase 'cultural alienation' has been used to describe the phenomenon. By today, there have been significant developments in the provision of education through the medium of the Maori language, particularly for the early years; teaching materials that are sensitive to the Maori culture have been developed (such as the inclusion of traditional Maori patterns in work on geometry); and the mathematics is set in contexts that reflect the natural experiences of the children. As a consequence, some of the initial concerns have begun to fade.

Similar experiences have been repeated worldwide. In attempts to modernise their methods of education, and under an historical influence as past colonies of European empires, mathematical ideas and resources were imported from the 'civilised' West during the 1960s and 1970s. As a consequence, mathematics was taught through the language of the country of the 'conqueror' and set within that country's cultural framework. It is hardly surprising that children in the villages of Nigeria were unable to identify with the exploits of *Janet and John* in an English middle-class suburb. The mathematics curriculum in these countries needed to reflect the local culture and to do so in ways that recognised the richness of the mathematical skills that already existed in those cultures.

In our daily use of mathematics we use words and phrases such as 'four', 'square', 'more than', 'equal to' and so on, whose meanings have been commonly agreed through their everyday use, be it on a back street in Chennai or in a pub in the Townhill area of Swansea. Those meanings have been informed by our early education, by the media and by our personal experiences with our families and friends. We bring our own interpretations to the table as 'meaning' is hammered out. We are all different and we all put a different spin on what we're taught. We all create our own mathematics rather than doing someone else's.

This happens at the level of whole communities as well as at a personal level, as can be seen most clearly in those communities that are relatively cut off from outside influences. For example, in the language of the Yancos, one of the Indian tribes living in the Amazon basin in South America, the word for the number 3 is pronounced as 'poettarrarorincoaroae'. The tribe's names for numbers are confined to those for 'one', 'two', 'three' and 'many' – the tribe hasn't needed to develop a more sophisticated method of counting in order to meet its daily needs. By contrast, think of the monosyllabic words used by western languages for the number 3: 'three', 'tri', 'trois', 'drei', and so on. In the event that the cultural setting of the Yancos were to change and that they found a need to use a more extensive numbering system, it is likely that their words for numbers would need to be extended and that those words that are already part of their vocabulary would be simplified.

As another example, the experience of many of the indigenous peoples of Australia has been based on their traditional life pattern of moving from place to place across a desert landscape that, to western eyes, looks unwelcoming and desolate. Were we to find ourselves in such a place we would immediately have a strong sense of being lost. That isn't the experience of the Aborigines. Their sense of place and of direction is highly developed. In many of the Aboriginal indigenous languages there are no words that correspond to our 'being lost', because 'being lost' isn't part of their experience. Neither are they familiar with phrases that correspond to 'turn to the left' or 'turn to the right' and they have no words that correspond to 'left' or 'right'. Rather, they have an extraordinary sensitivity to compass directions and can respond immediately, from an early age, to the equivalent of instructions such as 'turn to the south-west'. Speakers of the indigenous language Guugu Yimithirr – a language from which we have adopted the word 'kangaroo' – have an acute sense of direction. They instinctively know where 'north' is, in all weathers, day or night, be they standing still or on the move. Their environment has completely determined and enriched the way in which they understand geometrical ideas.

Doing mathematics is a creative act that parallels the creativity that we may normally be more ready to associate with the traditional arts. The poet hones her lines within a framework of her choice, and her creativity can be sharpened by the rules of rhythm and rhyme; similarly, the mathematician composes her mathematical verses within the constraints of the discipline. But there are close parallels between the two processes.

More often than not, the finished product of both the poet and the mathematician are presented in their final polished forms, all signs of sweat and hard work having been swept aside leaving the poem or the proof in pristine condition. In contrast to poetry, this polishing-up process in mathematics can hide the cultural context and only helps to sustain a myth that mathematics is divorced from such contexts and stands aloof from the nitty-gritty of daily life. There is some truth in that, of course, but it only tells half the story. This sanitisation of the process is well captured by this personal view of what mathematics is about:

> Mathematics makes me think of a stainless steel wall – hard, cold, smooth, offering no handhold, all it does is glint back at me. Edge up to it, put your nose against it, it doesn't give anything back, you can't put a dent in it, it doesn't take your shape, it doesn't have any smell, all it does is make your nose sore. I like the shine of it – it does look smart and intelligent … but I resent its cold impenetrability, its supercilious glare.

Taken from an article entitled 'An experience with some able women who avoid mathematics', this quotation also tells us something about the cultural perspective of women in relation to mathematics, a topic that we shall explore later in the book.

The history of the development of the familiar equals sign '=' provides an example of this process of sanitising. The sign was invented by the Welsh Tudor mathematician, Robert Recorde, a native of Tenby (see chapter 4 for more on Recorde's life and work). Recorde introduced the sign almost as an afterthought in his book on algebra *The Whetstone of Witte* (1557). Using his dialogue method to develop mathematical ideas between the teacher and his pupil, Recorde reveals that he found the need to write 'is equalle to' every time he wanted to work with equations was too cumbersome. In order 'to auoide the tedious repetition of these woordes' he had himself been using a shorthand and now wanted to introduce that shorthand to the pupil, where it appears on the page as a long, straggly and somewhat cumbersome symbol:

The symbol wasn't an overnight hit and many writers continued to use a variety of other methods to express equality, such as using the symbol *æ*, from the Latin *aequalis*, and that as late as the beginning of the eighteenth century. It took many years for the short, tidy sign with which we are so familiar today to gain universal use. The symbol is now firmly part of the 'stainless steel wall' that is our visual perception of mathematics. The sweat and mental toil – the human endeavour – that went into creating and developing the symbol has been quietly forgotten.

The same experience applies to virtually every part of mathematics, from the development of our numbering system based on the familiar Hindu–Arabic numerals 1, 2, 3,… to the most modern mathematical ideas. Behind the symbol, behind the theorem, the creative process has been at work polishing and perfecting.

During the Québec conference a close relationship developed between the Maori group and a group of Inuit delegates from Canada's Northwest Territories. There were marked differences in the outward temperament and demeanour of the two groups: the Maori were confident and expressed their ideas and opinions fluently and readily; by contrast, the Inuit were restrained and unassuming. As the week went by, the two groups came to respect each other and to recognise that, despite their cultural differences, they shared the same objectives. To celebrate this new friendship a message was sent from the head of the Inuit tribe inviting the Maori to travel north after the conference to spend some days as guests of the Inuit before returning home. The invitation was warmly welcomed and the Maori made immediate arrangements to fly to meet their hosts. It turned out that the airline couldn't accommodate all of the Maori in one plane and that they would have to travel as two sub-groups on separate planes and at different times. Before finalising the arrangements the Maori informed their hosts of their intention, only to receive an immediate response from the Inuit head: 'These arrangements are not acceptable', he said. 'You must arrive as one group so that we can welcome and greet you all at the same time. I shall send a plane from north Canada to Québec that will bring you all back here together.'

In his address at the beginning of the conference the leader of the Maori choir could easily have added, 'Our culture greets your culture. Our mathematics greets your mathematics.'

Explorer's puzzle

An explorer walks ten miles due south, then ten miles due east followed by ten miles due north, and finds that she's back at her starting point.

Where on earth is she?

'NOTHING WILL COME OF NOTHING'

King Lear: Lear

NDIA IS A complex and bewildering mix of excellence and corruption, of affluence and poverty, of religion and superstition, of pride and servility. In 2009 in the chaotic city of Chennai (formerly Madras), I'd been booked into a luxurious hotel surrounded by a high wall that, so we were told, kept out the city's beggars. The wall also protected a small number of privately owned shops set within the gardens of the hotel. One of those shops was a unique bookshop owned by Mrs Patel, a middle-aged and highly cultured businesswoman. This one-room shop was packed to the rafters with books to the extent that it was impossible to gain entry into the shop itself. All business was conducted on the pavement outside the shop, where I sat on a low stool to discuss new and old books with Mrs Patel. In the event that a book needed to be found, Mrs Patel would send a servant – a diminutive, elderly man – to climb into the shop and to burrow into the mounds of books. He would eventually reappear with a wide grin, triumphantly clutching the required volume.

Mrs Patel was particularly interested to know if I had read *The Indian Clerk* and to ask my opinion of it. Written as a novel, the book is based on the life and work of the remarkable mathematician Srinivasa Ramanujan (1887–1920). Born to a poor family living in the province of Tamil Nadu in south India, Ramanujan acquired advanced skills in mathematics from an early age but did not have the means to go on to university to develop those skills further. While working as a clerk in a company based in Madras, he spent his leisure time doing his own research in mathematics. In 1912 he began to send his results to G. H. Hardy, acknowledged as the best mathematician of his period in Britain if not the world.

Hardy (no one referred to him by his first name) was a mathematics tutor at Trinity College, Cambridge, a college famous for its mathematicians since the days of Isaac Newton. Hardy was also gay, but such matters were not openly discussed at the time. Although the letters he received from Ramanujan were unsolicited and their content didn't conform to standard mathematical conventions, Hardy was sufficiently astute to recognise that this hitherto unknown clerk had a very special talent. He arranged for Ramanujan to travel to Cambridge, leaving his family in Madras. Over a period of seven years Ramanujan and Hardy collaborated to produce cutting-edge research the like of which had never before been seen. Ramanujan was appointed Fellow of Trinity College and elected Fellow of the Royal Society, an honour that was highly unusual for such a young scholar. The story has a sad ending, however. Ramanujan never really managed to adjust his daily habits to be able to live comfortably in Cambridge and the college food was not to his liking. He suffered from a series of illnesses, including dietary deficiencies, and had to spend time in hospital before returning to his family in Madras where he died at the early age of thirty-two.

Mrs Patel was very conversant with *The Indian Clerk* and Ramanujan had the status of a hero in her eyes. During my stay in Chennai it became clear that Ramanujan's name was familiar to the population generally. On a large mural at one of Chennai's universities, equal status is given to Einstein and Ramanujan. Nevertheless, Mrs Patel was harsh in her criticism of *The Indian Clerk*. For her, the book unnecessarily cast a shadow on Ramanujan's good name by linking him with a gay person. She made it clear that no copies of the book would be on sale at her shop.

Part of the fascination with Ramanujan was his unique ability to devise formulae as if from thin air, as a magician pulls a rabbit from

Figure 1. *A large mural (approximately 8 metres by 5 metres), showing Einstein (top left) and Ramanujan (bottom right), on one of the main buildings of Loyola University College in Chennai.*

a hat. India has a long tradition of believing that mathematical truths are revealed in dreams to particular individuals during periods of deep meditation. On a visit to a modern bookshop during my stay in India I bought a book that lists examples of the fruits of such revelations. The book is essentially a recipe of formulae with no accompanying explanations – a classic example of *knowing how* without *understanding why*. In this tradition Ramanujan effectively had the status of a high priest. One of the challenges that he faced during his period in Cambridge was how to come to terms with the western world's view of doing mathematics and to reconcile this perspective with his previous tradition. By today, the 'western' view of mathematics forms the basis of what is taught in schools and colleges in India while traditional Indian mathematics retains an unofficial status side by side with that – one of the many paradoxes that give the country its unique character.

India can also justly claim to be the cradle of modern mathematics, and in particular the cradle of our modern method of counting. It was in India during the fifth century that the symbol '0' for the number 'zero' was first used. The great thinkers of the classical period – the Greeks and Romans – had refused to accept 'zero' or 'nothing' as a number at all. We can make sense of the notion that we can count three people, they argued, but what sense can there possibly be in asserting that the number of people in an empty room is 'zero'? The philosophers of ancient Greece argued long and hard over the question and came to the conclusion that 'zero' could not be regarded as a number in its own right. Although the Romans used numbers to plan and build their bridges, roads and cities they remained burdened by an unwieldy method of recording numbers that didn't include a symbol for the number 'zero'. We are familiar with Roman numerals such as XVI (for 16) but no Roman numerals include a symbol for 'zero'.

Roman numerals puzzle

In Roman numerals 2016 is MMXVI, which has five letters.

Which year, so far, has the most letters?

The recognition of 'zero' by Indian mathematicians as a number in its own right was a major advance. It allows us to write a number such as 306, and to interpret it as three hundreds, zero tens and six units: the existence of the symbol '0' makes it possible for us to create the composite 306. Each numeral in the written number signifies a value related to its position, so that the '3' in 306 stands for three hundreds, whereas the '3' in 603 would stand for 3 units. By today, we take such things for granted and label this as the decimal method of counting, but those first few steps by Indian mathematicians in the fifth century were truly revolutionary.

This revolution didn't reach Europe overnight, of course. Ideas born in India were transported over many years by traders travelling west towards Egypt, where they were adopted and refined by Arab scholars. The cultural journey took these ideas across North Africa and onwards to Spain, and on to Italy and the rest of Europe. A journey of some 6,000 miles over a period of 750 years – eight miles a year on average. Because of their origins in India and their subsequent journey to Europe via the Middle East, the numbers are commonly referred to as being written using 'Hindu–Arabic' numerals and it is this system that has been adopted the world over to write numbers. We can see a number such as 275 in a sentence of English, Welsh, Arabic, Russian or Chinese: Hindu–Arabic numerals are the lingua franca of the mathematical world.

It was the talented Italian Leonardo Fibonacci (*c*.1170–*c*.1250) who first drew the attention of western scholars to the new numbering method in his book *Liber abaci*, published in 1202. Even then the Roman method of writing numbers continued to hold sway and the process of replacing it was long and arduous. It is likely that Hindu–Arabic numerals first reached Britain towards the end of the fourteenth century. One of the earliest Welsh documents that uses these new numbers is a manuscript written by the poet and scholar Gutun Owain, otherwise known as Gruffydd ap Huw ab Owain (*fl.* 1460–1500). Written in 1488 or 1489, the manuscript is mainly devoted to discussing matters relating to astrology and medicine. Linked to this discussion, Gutun Owain uses Hindu–Arabic numerals when noting church festival dates on his calendar (Plate 1).

When writing *The Grounde of Artes* in 1543 – the first original book on arithmetic to be published in English – some three hundred years after Fibonacci, Robert Recorde (see chapter 4) was careful to begin by introducing the 'modern' Hindu–Arabic numerals, linking them to the more familiar Roman numerals. The latter continue to be used to this day, in such contexts as on some clock faces (possibly to attract a higher price tag) or when noting the copyright date on a film or a television programme.

The seeds for developing our modern Hindu–Arabic numerals were sown in India in the fifth century; the numerals reached

Europe some seven hundred years later in about 1200; the first evidence that we have of them being used in Wales dates to around 1500; and remnants of the older Roman system persist to this day, five hundred years after that. Such is progress!

Our numbers are not gifts handed down to us from the gods: the symbols that we use have been created by ourselves and have been combined by us to form the sophisticated system with which we are so familiar today. An entirely different matter is how languages have developed to enable us to discuss those numbers and to put them into words. As we shall see later in this book, there are notable variations between languages, not least between the language forms used in Welsh and English, two languages that live cheek by jowl but whose histories of number differ significantly.

SETTING THE RECORDE STRAIGHT

W E ALL NEED our heroes, both to admire and to emulate. As a youngster, my cricketing hero was the spin bowler Tony Lock who played for England in the 1950s and 1960s. Like me, he bowled with his left hand but batted with his right. He was also a fabulous fielder, particularly at leg slip, and I would spend hours in the backyard at home practising catching skills in a (vain) attempt to emulate Lock's brilliance.

Following a career in mathematics education I naturally looked for heroes there as well. My clear favourite is Robert Recorde (1512?–58), the Welshman from Tenby best known for having invented the equals sign '='. He's a hero for me, not so much for that, but because he was the first in Britain to think seriously about the challenge of learning and teaching mathematics, about how to make mathematical knowledge and skills available to a wide population. He was, in essence, our first maths teacher. Recorde was the first to write original texts in English setting out the principles of arithmetic and algebra, and in so doing, yes, he was the first to use the sign of equality '='. He was also the first to interpret the classical geometry of Euclid for ordinary people. In the tradition of tragic heroes, Recorde died defending his principles. Mathematics does not deceive and Recorde himself was unwilling to be party to any form of deception. He paid the ultimate price for taking such a stand.

I began to take an interest in Robert Recorde's life and work in the early 1970s when I joined an association of lecturers in Wales specialising in mathematics education – those with responsibility for training prospective primary and secondary teachers. Why on earth had I not heard of Robert Recorde previously? I'd studied A-level Mathematics and had graduated in the subject, but no one had mentioned his name, let alone referred to his work. Shouldn't all schoolchildren know about this man of genius?

Quite naturally, Robert Recorde is acclaimed as a hero by the association of lecturers, and the members often refer to him playfully, but respectfully, as 'Dai Equals'. The year 2008 marked the 450th anniversary of Recorde's death and the association took that opportunity to hold a special conference to celebrate his life and work. The event was a resounding success and drew together international experts in the history of mathematics. What was so special about Recorde and why does he deserve a place in the pantheon of Welsh heroes?

Modern-day Tenby is a small, bustling town, particularly during the summer months when visitors arrive in droves to enjoy its golden sands and the temperate climate that characterises south Pembrokeshire. Once there, they cannot but notice the many historical landmarks in the town: the medieval walls that defend the centre of the town, the ancient church of St Mary's, and the Tudor house that once belonged to a local tradesman and which has been restored to its original condition.

During the summer of 2010 Tenby's narrow streets were bright with flags declaring to all and sundry that the town was that year celebrating both the 800th anniversary of the establishment of St Mary's Church and the 500th anniversary of the birth of its most famous son, Robert Recorde, albeit that it's more probable that Recorde was born in 1512 rather than 1510. Over the summer months the Tenby Museum and Art Gallery housed an exhibition of the work of over fifty artists who had been inspired by the life and work of Recorde (see Plate 2 for one such example). The anniversary was also celebrated in a memorial service held at St Mary's.

Figure 2. *The bust of Robert Recorde in St Mary's Church, Tenby, based on a painting originally thought to be of Recorde and now housed in the Faculty of Mathematics at the University of Cambridge. It has been established that the painting is by a Dutch artist and actually dates from the first half of the seventeenth century. This is certainly not Recorde. However, the image continues to be used today to represent Recorde in the absence of anything more authentic.*

Tenby was a small, bustling town back in 1500 as well. Remotely situated and difficult to reach by road, it was a popular port visited by ships from far and near, its harbour attracting traders and merchants of a wide range of goods. The success of the town depended on the trading skills of its businessmen. The port officials were also busy collecting taxes from the visiting ships. The common people, referred to by Recorde as 'the vnlearned sorte', needed a wide range of mathematical skills to be able to go about their business, including keeping accounts and ledgers, exchanging currencies, and weighing and measuring using a bewildering mixture of units.

One of the immigrants to Tenby during the fifteenth century was Roger Recorde, a merchant from Kent. Roger's only son, Thomas Recorde, was also a merchant in the town. Thomas married Ros Johns, the daughter of a Welsh family from Machynlleth in mid Wales. Robert Recorde was the younger of two sons of Thomas and Ros. Little is known of Robert's early life in Tenby but he was clearly influenced by the comings and goings in the busy port of Tenby and the extensive mathematical skills used by the tradesmen and craftsmen of the town. He probably attended a small school held in St Mary's Church and managed by one means or another to secure a place at Oxford. After graduating there, he went on to Cambridge where he qualified as a doctor, thus enabling him to practise medicine.

Recorde became known at Court and was appointed at various times to oversee the royal mints at Bristol, London and Dublin and to supervise silver-mining operations in Ireland. He augmented his salary from these sources by continuing to undertake some medical work and to give private lessons on mathematics. He was a conscientious worker but was politically naïve. He became embroiled in a long-running dispute with Sir William Herbert, later the Earl of Pembroke, whom he accused of malfeasance, involving the stealing of Crown property and diverting profits made by the mints for his own benefit, so depriving the Crown of its due income. Pembroke, who had previously accused Recorde of treason, now sued him for libel and claimed damages. Being a powerful member of the Privy Council, Pembroke had the upper hand and Recorde was unable to secure a fair trial. He was duly found guilty and sentenced to pay

Pembroke £1000 damages with costs. Pembroke did not press for an immediate payment but, in late 1557 or early 1558, Recorde was being held in the King's Bench prison at Southwark, south London, while further enquiries were being made into his affairs. One can only speculate as to how such a situation arose and at whose instigation but it was there that, within a few months, he contracted a fatal disease and died in tragic circumstances.

In 1570, twelve years after Recorde's death and following a period of lobbying by Recorde's nephew, the accounts of the mints and the mining operations prepared by Recorde were re-audited and it was confirmed that the government owed Recorde money for his services in Ireland. Queen Elizabeth I ordered that Recorde's family be compensated and the Crown gifted leasehold land in Tenby and elsewhere to the care of Recorde's nephew. It was, of course, too late to recompense Recorde himself, but this arrangement secured the wellbeing of his family in Tenby.

It is quite remarkable that, under these pressures, Recorde was also able to write and publish his series of books, not only on mathematics, but also on medicine and astronomy. These publications were his main achievement. The mathematical texts laid the foundations for the teaching of mathematics in English, setting a standard to be emulated for generations to come. When discussing numbers and shapes with pupils today, every teacher and every parent, whether or not they realise it, relies on the principles of good teaching exemplified by Recorde. It is therefore not surprising that he has hero status among mathematics educators. Neither is it surprising that he is a hero in his home town of Tenby. The recognition accorded to him in wider circles, including in Wales, has taken longer to blossom, but that is also changing swiftly.

Robert Recorde was a Protestant of strong conviction. He also understood the scientific ideas of his time. In particular, he supported Copernicus' view that the earth orbits the sun rather than believing that the earth is at the centre of the universe, as claimed by the Church of Rome. It was a dangerous period in which to hold such heretical anti-church views, particularly during the reign, from 1553 to 1558, of the Catholic Queen Mary (Bloody Mary). This did not stop Recorde from expressing his opinion but he chose to do so somewhat indirectly. His discussion of the question in his book on astronomy, *The Castle of Knowledge* (1556), is set out as a conversation between a master and a scholar. The scholar suggests that there is no merit in Copernicus' views before the master chides him for venturing an opinion on something so complex: 'You are too young to be a good iudge in so great a matter.' The implication is clear, nonetheless, although historians are not unanimous in their view as to whether or not Recorde was in fact the first British Copernican.

Recorde's command of literary English is extraordinary; his prose flows easily and rhythmically despite being somewhat strange to our modern ears, particularly since spelling and punctuation hadn't been standardised at that time, but the meaning becomes clear with minimal effort. In the box (below) he charges his readers to question other people's views rigorously (Recorde's original words are on the left and a modern rewording is set out on the right).

Yet muste you and all men take heed, that … in al mennes workes, you be not abused by their autoritye, but euermore attend to their reasons, and examine them well, euer regarding more what is saide, and how it is proued, then who saieth it: for autoritie often times deceaueth many menne.	*When considering the views of others take care not to be put off by their status and authority. You must weigh up their evidence carefully rather than relying only on whose arguments they are, because people in authority can often deceive us.*

In Recorde's time this advice could be interpreted at different levels as it could be applied to members of the political class, as well as to the Church of Rome. The words can be seen today on a memorial slate that was unveiled in 2001 in the Department of Computing at Swansea University – they are as pertinent to the current generation of students as they were to Recorde's readers.

Figure 3. *A woodcut from the title page of* The Ground of Artes *showing merchants using two different methods of calculation: moving counters as on an abacus on the left, and writing the numbers as a sum on the right.* (Courtesy of TGR Renascent Books)

But Recorde's books on mathematics form the highpoint of his work: his first on arithmetic, *The Ground of Artes* (1543 and 1552); his second on geometry, *The Pathway to Knowledg* (1551); and his last book on algebra, *The Whetstone of Witte* (1557). His aim throughout was to present mathematics to ordinary people and to do so in a language that they could follow with relative ease.

These were the first original books ever to be published on mathematics in English, rather than following the tradition of writing in Latin or Greek for a handful of scholars. Recorde interpreted the classics for ordinary people. In his book on geometry, for example, he interprets the work of the Greek mathematician Euclid (dating from about 300 BC) for the first time in English, his audience comprising both ordinary people and scholars. He includes those who work the land and need elementary geometry to measure their fields in order to grow their crops, to plant their hedges and to dig their ditches; he also includes craftsmen of all kinds – the carpenter, the sculptor, the stone mason, the cobbler, the tailor – who use simple ideas in geometry in their daily measuring and designing. It proved to be a popular publication and ran to two further editions after Recorde's death, in 1574 and 1602. It is difficult to overemphasise Recorde's achievements; this Welshman from Tenby was the first to open the eyes of ordinary English-speakers to the wealth of Greek mathematics.

In his books on arithmetic and algebra Recorde again uses a dialogue style between teacher and pupil, referred to by Recorde as master and scholar:

I haue wrytten in the fourme of a dyaloge, bycause I iudge that to be the easyest waye of enstruction, when the scholer may aske euery doubte orderly, and the mayster may answere to his question playnly.	*I have written the book as a dialogue because I believe that to be the easiest way to teach, allowing the pupil to ask questions for clarification and the teacher to provide clear answers to those questions.*

At one point in the dialogue the pupil loses his patience and asks the teacher for the answer without wanting to understand *why* that answer is correct. We can understand his frustration: he's under pressure to finish his task, and wants the answer without having to go to the trouble of understanding why it works – a situation that is very familiar to both teachers and parents today. But the teacher wants none of it:

Yea but you muste proue your selfe to doe some thynges that you were neuer taught, or else you shall not be able to do any more then you were taught, that were rather to learne by rote (as they call it) then by reason.	*Yes, but you must be able to show that you can do some things that I haven't taught you. Otherwise, you will be reduced to having to repeat methods mechanically rather than using your ability to reason.*

> Learning *by rote* means mechanical repetition with no understanding
> – *knowing how* rather than *understanding why*.

In this example the pupil is content on *knowing how*, but the teacher insists on him *understanding why*. Mastering mathematics is not achieved by learning something mechanically, even in Recorde's time.

From beginning to end, Recorde's books reveal a person who thought through with great care how to teach mathematics, what needed to be explained to the pupil, how to retain his interest, how to correct him, how to fire his imagination and how, also, to use humour from time to time as part of the teaching process. His book on arithmetic established a pattern that was not bettered for more than three hundred years. Recorde himself expanded the scope of the book in a second edition. During the 150 years following his death a series of editors adapted, expanded and reprinted the book more than thirty times.

Despite the importance and influence of his books on mathematics Recorde is best known today for just one thing, for it was he who devised the symbol '=' to mean 'equals'. The symbol is so familiar to us by now that it is easy to gloss over its power and significance. Recorde was used to expressing equality by writing the words 'is equalle to', but he recognised the awkwardness of doing that each time and that it would be far easier to replace the words with a nice tidy symbol. He introduced the new symbol in his book on algebra, choosing two parallel lines to represent the idea 'bicause noe. 2. thynges, can be moare equalle'. The quotation is given in full overleaf, in the original typeface, with an adaptation into modern English beneath.

Some of the other mathematical symbols – those for addition and subtraction, for example – had already been introduced by German mathematicians, and Recorde was the first to use them

> 𝔄𝔫𝔡 𝔱𝔬 𝔞=
> 𝔲𝔬𝔦𝔡𝔢 𝔱𝔥𝔢 𝔱𝔢𝔡𝔦𝔬𝔲𝔰𝔢 𝔯𝔢𝔭𝔢𝔱𝔦𝔱𝔦𝔬𝔫 𝔬𝔣 𝔱𝔥𝔢𝔰𝔢 𝔴𝔬𝔬𝔷𝔡𝔢𝔰 : 𝔦𝔰 𝔢=
> 𝔮𝔲𝔞𝔩𝔩𝔢 𝔱𝔬 : 𝕵 𝔴𝔦𝔩𝔩 𝔰𝔢𝔱𝔱𝔢 𝔞𝔰 𝕵 𝔡𝔬𝔢 𝔬𝔣𝔱𝔢𝔫 𝔦𝔫 𝔴𝔬𝔬𝔷𝔨𝔢 𝔲𝔰𝔢, 𝔞
> 𝔭𝔞𝔦𝔯𝔢 𝔬𝔣 𝔭𝔞𝔯𝔞𝔩𝔩𝔢𝔩𝔢𝔰, 𝔬𝔷 𝕮𝔢𝔪𝔬𝔴𝔢 𝔩𝔦𝔫𝔢𝔰 𝔬𝔣 𝔬𝔫𝔢 𝔩𝔢𝔫𝔤𝔱𝔥𝔢,
> 𝔱𝔥𝔲𝔰 : ===== 𝔟𝔦𝔠𝔞𝔲𝔰𝔢 𝔫𝔬𝔢, 2, 𝔱𝔥𝔶𝔫𝔤𝔢𝔰, 𝔠𝔞𝔫 𝔟𝔢 𝔪𝔬𝔞𝔯𝔢
> 𝔢𝔮𝔲𝔞𝔩𝔩𝔢,

In order to avoid having to write 'is equal to' every time, I shall use a symbol that I often use in my own work, a pair of parallel lines of the same length (Gemowe lines) like this = because no two things can be more equal than that.

The term 'Gemowe lines' in this quotation emphasises that the lines, like twins, are exactly the same. 'Gemowe' has the same root as Gemini, the star sign of the Twins.

Figure 4. *Robert Recorde introduces his readers to the equals sign in* The Whetstone of Witte *(1557).* (Courtesy of TGR Renascent Books)

in books written in English. His sign for equality now allowed equations to be written in a more compact form. An equation such as $3 + 5 = 8$ is so familiar to us today that it is difficult fully to appreciate the enormity of the conceptual leaps that were made to reach this simplified form.

Recorde's book on algebra was published in 1557 while he was embroiled in his legal difficulties with the Earl of Pembroke. At the very end of the book the teacher and pupil are discussing some technical details in the algebra and are completely absorbed in their work. Suddenly, and dramatically, there is furious knocking at the door. Court officials arrive to serve a writ on the teacher – Recorde himself, of course – that requires him to present himself at court. He has to set aside his mathematics in order to answer his accuser. The pupil is disconsolate and the book ends with this heartrending scene, set as an upturned triangle, the lines shortening as the end approaches:

> *Scholar:* My harte is so oppressed with pensifenes,
> by this sodaine vnquietnesse, that I can not expresse
> my grief. But I will praie, with all theim that
> loue honeste knowledge, that God of his
> mercie, will sone ende your troubles,
> and graunte you soche reste, as
> your trauell doeth merite.
> And al that loue lear-
> nyng: saie ther
> to. Amen.
> *Master:* Amen,
> and Amen.

In modern prose:

> Scholar: *My heart is so heavy with sadness*
> *at this sudden disturbance that I am unable to express*
> *my grief. But I pray with all those who have a love*
> *for knowledge that God, in his mercy,*
> *will soon dispel your problems*
> *and grant you the peace*
> *that your work deserves.*
> *And all those who love*
> *learning, say*
> *Amen to that.*
> Master: *Amen,*
> *and Amen.*

Within a matter of only a few months Recorde had died. This remarkable man had intended to publish more books but was not given the opportunity to realise his plans. Nevertheless, he had already achieved much and had done so under difficult political circumstances. By keeping to his principles and speaking plainly and honestly, he had crossed swords with his masters in the royal court. His wish had been to continue as a teacher, a communicator of mathematical ideas. That work came to an abrupt end, but its influence remains.

The Welshman Robert Recorde set the foundations of mathematics education in English. His aim was to present basic mathematical ideas to ordinary people – clerks, craftsmen, labourers, merchants, sailors, shopkeepers, teachers – as well as to scholars. In so doing, he also set out the basic principles of good teaching that have stood the test of time. A hero, if ever there was one!

In the absence of a university in Wales until late in the nineteenth century, Robert Recorde was drawn to the main learning centres in England – Oxford, Cambridge and London. His contribution to Welsh cultural life was thereby minimised. That culture has been slow to embrace science and mathematics. But Recorde's work was not without its followers in Wales, as we can see in the next chapter.

A weighty puzzle

This is one of the problems posed by the teacher to his pupil in Robert Recorde's *The Ground of Artes:*

If the caryage of 100 pound weyghte 30 myles, do coste 12 d. how much wyll the caryage of 500 weyghte coste, beynge caryed 100 myles?

In modern English:
The cost of carrying a weight of 100 pounds a distance of 30 miles is 12 pence. What is the cost of carrying a weight of 500 pounds a distance of 100 miles?

5 'NEITHER A BORROWER NOR A LENDER BE'

Hamlet: Polonius

MYR LOOKED smug. He had just done a whole page of sums and finished his work well before the end of the morning lesson. Job done! Emyr was nine years old and his teacher had given exercises to the class to make sure that they could all remember how to do subtraction sums. I sat with Emyr and looked at the first two sums on his page:

$$47 \qquad\qquad 31$$
$$\underline{-23} \qquad\qquad \underline{-18}$$
$$24 \qquad\qquad 27$$

His first answer is correct, of course, but something strange has happened in the second sum. How did Emyr get the answer '27'? It's likely that he took 1 from 8 in the units column and then took 1 from 3 in the tens column, without realising that there was something not quite right with that. After all, wasn't he supposed to be doing 'take-aways' that morning? What could possibly be wrong with that?

I put the sum to one side and asked Emyr a different question. 'Imagine', I said, 'that you've got thirty-one (31) sweets and you give eighteen (18) of them to me. How many would you have left?' Emyr thought for a while, his eyes fixed on the far corner of the room. After a while, he replied, 'Thirteen (13).' 'That's interesting',

I said. 'How did you get that?' What, I wondered, had the far corner revealed to Emyr? 'Well', he said confidently, 'eighteen (18) and two (2) makes twenty (20), another ten (10) makes thirty (30), and one (1) more makes thirty-one (31). So that's it – two (2), ten (10) and one (1) make thirteen (13).' Emyr had seen it! Whatever was lurking in that far corner had helped him to reach into the recesses of his mind and to work it out. Rather than taking 18 away from 31 he had asked himself how many more is 31 than 18 and had done that calculation by working out how many he had to add to 18 in order to reach 31. That's how he got the 2, the 10 and the 1, and put them together to make 13.

The question had given Emyr the opportunity to think creatively and to show that he *understood why.* But that raised a new dilemma. How could Emyr reconcile the answer that he'd given in his exercise book (27) with the answer that had been inspired by the corner of the room (13)? He had already written 31 – 18 = 27 in his exercise book but had given me the answer 13 orally. I decided to confront Emyr with this dilemma: 'You've written twenty-seven (27) in your book but then you gave me the answer thirteen (13). Which of them is right?' After further thought, Emyr came to a reasoned conclusion that settled the whole thing: 'Well', he said, *'both* answers are right!' Emyr had separated the two things in his mind. The answer 27 was 'right' because he had followed the 'rules' that he had learned from his teacher – wasn't that why he spent an hour each morning in the maths lesson learning rule after rule so that he could get answers to add-ups, take-aways, times sums and gusinters (Emyr's term for division sums – 3 'gusinter' 12)? At the same time 13 was 'right' as the answer to a practical question. The former was what you did in maths lessons; the latter was what you did in the real world outside school.

Generations of teachers have sweated long and hard in their efforts to introduce children to the mysteries of subtraction sums. We can use Emyr's sum to explore the problem. How do we tackle a sum like this?

$$3|$$
$$-\underline{18}$$

The standard method starts by looking at the units (on the right) and tries to take '8' away from '1'. It appears that we can't do that, and children have been taught to say 'one take away eight, you can't'. Emyr had seen the problem coming and had changed the order to 'eight take away one'. That, of course, was how he got '7' as his answer.

These days, children are taught to take a closer look at the two numbers, 31 and 18, and to interpret them as:

$$31 = 30 + 1$$
$$18 = 10 + 8$$

After seeing that you can't 'take 8 from 1' it's then a short step to rewrite 31 as 20 + 11:

$$31 = 20 + 11$$
$$18 = 10 + 8$$

We can now see that we can take '8' from '11' to give '3' and then to move on to the tens by taking '10' from '20' to leave '10'. The take-away sum now looks like this:

$$
\begin{array}{r}
31 = 20 + 11 \\
-18 = 10 + 8 \\
\hline
10 + 3
\end{array}
$$

It follows that the answer is '10 + 3', or 13.

If you're familiar with this method of subtraction you may also remember that the technical term for the method is 'decomposition', because we 'decompose' 31 so as to write it as 20 + 11. On paper the process may be recorded as something like this:

$$\begin{array}{r} {}^2\not{3}{}^{\scriptscriptstyle|}1 \\ -18 \\ \hline 13 \end{array}$$

It's also possible that this method is completely strange to you but that you have a dim and distant memory of something quite different. In broad terms, if you were a primary-school pupil during or after the 1980s then it's likely that you were introduced to the decomposition method. If, like me, you were at school before then, you were probably introduced to the mysteries of a method that is much more difficult to explain but looks tidier on paper – a method that concentrates on *knowing how* rather than *understanding why*. You may also remember using phrases like 'borrow and pay back' when you did your take-aways.

A common experience is to see parents or grandparents who were reared on the method of 'borrowing and paying back' trying to help their children or grandchildren who are using the 'decomposition' method – and everybody being completely confused. Using the 'borrow and pay back' method, Emyr's sum would look like this:

$$\begin{array}{r} 3{}^{\scriptscriptstyle|}1 \\ -1\,\underline{8} \\ \hline 13 \end{array} \qquad \text{or} \qquad \begin{array}{r} 3{}^{\scriptscriptstyle|}1 \\ -{}^2\not{1}8 \\ \hline 13 \end{array}$$

When learning this method children were taught to say, 'One (1) take away eight (8), you can't, so borrow one (1) to the top and pay back one (1) to the bottom. Eleven (11) take away eight (8) is three (3); three (3) take away two (2) is one (1). Answer, thirteen (13).' If you didn't follow that, don't worry, neither can most other people. It's complete nonsense and a perfect example of mechanical repetition, of *knowing how* but not *understanding why*. Yes, the answer is correct and if that's your only objective then the method of 'borrowing and paying back' suits your needs. But if you're trying to understand what you're doing, then this method is not for you. Worse than that, the use of this method encouraged children to believe that you weren't supposed to understand what you were doing in a sums lesson, that repeating rules like a parrot was the aim and that you should settle for that. Doing sums was a means of keeping you quiet, not something to understand and certainly not something to enjoy.

It is, however, perfectly possible to explain why the 'borrow and pay back' method works and why it gives the correct answers but the explanation is far too difficult for most young children to understand. It is, in fact, too difficult for many adults to grasp. Generations of teachers have taught the method to their pupils without being able to offer them any form of explanation or justification. They had to be content with teaching them *how* to do the sum mechanically, without being able to explain to them *why* it worked. Very few primary-school teachers understood the method themselves, a fact that should not cause any surprise. Nevertheless, teachers had to do their level best to help their pupils to perform the method by drilling them to remember the words they had to repeat mechanically every time.

I have a vivid recollection of a visit I made to a small rural school in the early 1980s. All the infants were taught in one class and there was a piano in the corner of the room. 'Modern' methods were yet to reach this school and the teacher was trying to help the 'top infants' to remember how to do take-aways. 'Do you remember, children? You have to "borrow one" here, don't you?', adding, 'and borrow it from behind the piano.' I didn't have the heart to ask her how the children were meant to cope with take-aways when the piano wasn't there! In some primary schools in the coal-mining valleys of south Wales children were helped to remember 'borrow and pay back' by reference to the image of 'one up the winder and one down the pit'. The phrase used in some Catholic schools was 'one for heaven and one for hell'. The phrase 'borrow and pay back' has no meaning. It's an ancient phrase, conjured up in the dim and distant past, but nonsense is nonsense, ancient or not.

The 'borrow and pay back' method survived the test of time because it is marginally tidier on paper than 'decomposition', and that was a strong argument in its favour when keeping written accounts was the norm in business and commerce, in banks and post offices. Rows of clerks were employed to undertake such tasks. By today, computers and calculators have removed the need for such laborious work. Children still need to be able to subtract, of course, but the standard method used in schools is 'decomposition' and teachers are now able to undertake their professional responsibility to explain both how and why it works.

Robert Recorde had thought long and hard in 1543 about the best way to introduce subtraction in his book on arithmetic, *The Ground of Artes*. The 'borrow and pay back' method was well established, even by then, but Recorde recognised that it wasn't easy to explain it. Ultimately, he chose to duck the issue. After the teacher shows the rule (the *knowing how*) to his pupil, the teacher then adds that his intention is to explain it (the *understanding why*) at another time. Recorde is clearly not comfortable with this approach, because his stated aim was to ensure that the pupil could 'se the reason in euery thynge'. Although he promised to return to the matter at a later stage, he was unable to keep to his word and his mind was diverted to other arithmetical matters. Recorde would certainly have been able to sympathise with the armies of teachers who, over the five hundred years or so since that period, have also struggled to teach subtraction to their own pupils.

Ducking the issue has been common practice by the authors of other books on arithmetic since the days of Robert Recorde. The Welsh author John Thomas (1757–1835) was a notable exception. Born in the village of Llannor near Pwllheli in north-west Wales, John Thomas undertook a variety of occupations during his life, spending time as a weaver, a sailor and a schoolmaster as well as a collector of taxes in Liverpool. In 1795 he published a book in Welsh in four parts under the title *Annerch i Ieuengtyd Cymru* (roughly translatable as *An Address to the Youth of Wales*). The second part of the book is a text on arithmetic. John Thomas had been reading one of the later editions of *The Grounde of Artes*: the style and content of the *Address* is strongly influenced by Recorde's work. John Thomas uses some of Recorde's examples and follows Recorde's style of introducing and discussing mathematical ideas using a dialogue between a teacher and his pupil, referring to the teacher as 'Philo' and to the pupil as 'Tyro'. Philo is an abbreviation of Philomathes, a Greek name for one who loves knowledge, and Tyro derives from the Latin *tiro*, a young soldier undergoing training.

In John Thomas's book, the teacher helps the pupil to gain an understanding of each topic and encourages him to spend time practising his newly acquired skills so as to deepen his understanding. He offers the advice that 'an hour spent practising is more valuable than a day spent talking'. The quandary of how to teach subtraction sums confronted John Thomas as it had previously confronted Robert Recorde. After Philo introduces the method based on 'borrow and pay back', Tyro confesses that he is utterly confused and asks Philo for more help: 'I can make little sense of this business of borrowing, please show me some more examples to make it clearer' (my translation).

Philo is more than happy to help out, and leads Tyro step by step through the subtraction sum 834 – 679. Tyro finds this very helpful but continues to be uneasy regarding the notion of 'borrow and pay back', and makes a comment to the effect that it would be so much better to use a method based on the notion of 'decomposition' (although he doesn't use that particular term) as that would be easier to understand. Between the lines of the dialogue we sense that the teacher agrees with Tyro and that he can only respond by saying that, for better or worse, the method based on 'borrow and pay back' is the one that has gained currency. It's this method, says Philo, with an air of sadness, that's commonly used, but the teacher is clearly not satisfied with his own explanation. It's precisely this frustration that primary-school teachers have faced over the years – having to settle for a method that they couldn't explain.

John Thomas had benefited from the work of another Welshman, Robert Recorde, who had shown the way some 250 years previously. John Thomas was intent on presenting mathematical ideas to his countrymen in their own language – Welsh – and did his level best to offer full explanations. Emyr would have been in his element as a pupil in his class.

The tension between *knowing how* and *understanding why* is an old and recurrent theme in mathematics education. Teachers in Victorian-age schools aimed to knock sums into the heads of working-class pupils. Pupils in public schools, meanwhile, were introduced to the fine art of Euclidean reasoning in their geometry lessons. *Knowing how* and obeying instructions was the fate of the working classes; *understanding why* and learning how to reason was the aim for the ruling classes.

Thomas Gradgrind was the Victorian teacher in Charles Dickens's *Hard Times*, his name being sufficient to suggest the narrowness of the educational experience that he provided for working-class children: 'Now, what I want is, facts. Teach these boys and girls nothing but facts. Facts alone are wanted in life. Plant nothing else, and root out everything else.' More than a hundred years later, Rhodes Boyson, then a junior minister for education in Margaret Thatcher's government, was being interviewed on television. In response to a question about the National Curriculum, that which all children in state schools are required to learn, Boyson said, 'Two twos are four, you don't ask why, it just is.' For Rhodes Boyson, a twentieth-century Gradgrind, learning the fact, rather than understanding it, was paramount. The National Curriculum isn't compulsory in public schools.

In a display of educational radicalism, Robert Recorde had introduced mathematical ideas to ordinary English-speaking people. Two hundred and fifty years later, John Thomas extended this radicalism by addressing the needs of working-class Welsh speakers.

Puzzle of the 'Double Rule of Three'

In 1969 the popular Welsh duo Tony and Aloma recorded a song in Welsh about a small village school. The words of the song include a remarkable verse about the teaching of 'numeration' that culminates in the 'Double Rule of Three'.

What was the 'Double Rule of Three'?

AMAZING MAYANS

HERE ARE TWO counting systems in Welsh: a modern decimal system, based on counting in tens, and a traditional vigesimal system, based on counting in twenties. The details of both systems are set out in the Appendix. For example, the number eighty-six (86) is expressed straightforwardly in the modern decimal system as *wyth deg chwech* (eight tens six), while in the traditional vigesimal system, it appears as *chwech a phedwar ugain* (six and four twenties). A complication in the vigesimal system is the use of ten and fifteen as stop-off points so that, for example, thirteen (13) is *tri ar ddeg* (three on ten) and sixteen is *un ar bymtheg* (one on fifteen). An additional curiosity is the use of *deunaw* (two nines, 2×9) for eighteen (18), where, for consistency, we might have expected *tri ar bymtheg* (three on fifteen). While we know how the decimal system of *writing* numbers developed, it is far less clear as to where the vigesimal system of *speaking* those numbers had its origins. Why, for example, was *deunaw* introduced? We may suspect that there may have been agricultural reasons, linked possibly with a practice of stacking sheaves of corn in nines so that a *deunaw* could refer to two such stacks, but there is no clear evidence for this or any other reason. It remains a curiosity, but one which is still used in modern Welsh in particular contexts. For most practical purposes, however, the vigesimal system has been displaced by the decimal system.

Towards the end of the 1980s I was contacted by Annie MacDonald on behalf of schools on the Western Isles of Scotland for advice regarding the counting system in Scottish Gaelic. Gaelic-medium primary schools in Scotland were relatively new – the system of Welsh-medium schools in Wales had got off the ground back in the late 1940s and 1950s. Annie MacDonald's concern was how to teach number to young children, given that Gaelic's traditional way of expressing numbers was based on a vigesimal system, and the language hadn't developed a parallel decimal system. It was only then that I fully realised that the traditional way of counting in twenties – the vigesimal system – wasn't unique to Welsh but was also common to the other Celtic languages: Breton, Cornish, Irish Gaelic, Scottish Gaelic and Manx (the variant of Gaelic spoken on the Isle of Man). This feature of the Celtic family of languages is further enriched by variations between the languages. For example, while *deunaw* (two nines) is the Welsh word for eighteen (18), the equivalent word in Breton is *triwec'h* (three sixes, 3 × 6).

Traces of a vigesimal system are also evident in some other European languages, including the French *quatre vingts*, or four twenties, for eighty, and in the English expression from the St James Bible, 'the days of our years are threescore years and ten', being 3 × 20 + 10, or 70. Beyond Europe, the Mayan civilization developed one of the most striking examples of the use of a vigesimal system of counting. The Mayans occupied parts of Central America between about AD 600 and AD 1200, covering an area that roughly corresponds to what is now southern Mexico, Guatemala, Belize and parts of El Salvador and Honduras.

The Mayan method of counting was based on three elements. Firstly, they devised a symbol for zero, paralleling the development in India – a remarkable feat, particularly in a part of the world that was insulated from outside influences. The Mayans used a shell to represent zero, and their written symbol for zero was based on the shape of a shell:

Secondly, the Mayans used a sub-system based on counting in fives. In this system a small pebble represented the number 1 and a stick represented the number 5. The pattern below, using a pebble and three sticks, would then represent the number 16, its form bearing a striking physical resemblance to the Welsh *un ar bymtheg* (one on fifteen):

$$
\begin{array}{r}
1 \\
5 \\
5 \\
\underline{5} \\
16
\end{array}
$$

Thirdly, the Mayans extended their counting to higher numbers by adopting a vigesimal system and applying it consistently and universally. In order to show the number 36, for example, they split it up as a twenty to which you added sixteen and set this out as a column. The first 'entry' in the column corresponds to the number of twenties (one pebble in this case, representing 20) and the second 'entry' to the remaining 16:

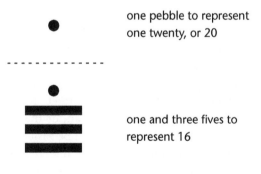

Total: 20 + 16 = 36

Keeping to the same pattern, this is how the Mayans represented the number 80, as four twenties in the first grouping of pebbles, followed by a shell to show that there is nothing to be added:

four pebbles to represent
four twenties, or 80

 a shell to represent zero
units, or 0

Total: 80 + 0 = 80

Using only a few shells, pebbles and sticks, the Mayans performed calculations swiftly and accurately using their vigesimal system. Their methods were streets ahead of those being used in Europe during the same period. The Mayans also extended the method to form a basis for their sophisticated calendar system. Continuing this system to our own period, the numbers revealed that one complete cycle of the Mayan calendar finished on 21 December 2012. On that basis some prophets of doom predicted that the world would come to an end on that day to be reborn the following morning. There is no evidence that such an event occurred, although there was no shortage of people who took precautions against such a cataclysm.

On visiting Mexico City in 2004 I was intrigued to see how modern Mexicans sell the history of their numbers to tourists. And yes, this tourist just had to buy a date depicted as one on the Mayan calendar. The image (see Plate 3) uses a series of glyphs to represent the birth date of our son, 18 May 1976, a record that is marginally more colourful than his official birth certificate.

There is no evidence that the Mayan vigesimal system in any way influenced the development of the Celtic vigesimal system the other side of the Atlantic Ocean, or that the Celts had influenced the Mayans. Rather, it would be easier to believe that there may have been similarities in the underlying causes. One theory proposed by an expert on Mayan culture is that it would have been natural for barefooted people living in a hot climate to have used all their fingers and toes for counting and that that may have been a factor. Be that as it may, one wonders whether the same argument would be equally persuasive if applied to medieval Europe.

We can at least conclude that the traditional Welsh vigesimal system is neither odd nor unnatural. It's embedded within the wider Celtic group of languages and was also central to the Mayan tradition, which applied it to all their numbers, however large. By contrast, the use of the traditional vigesimal system in Welsh was more restricted. Building on the Welsh word for twenty, *ugain*, the most commonly used multiples of twenty in the traditional Welsh system are *deugain* (two twenties – 40), *trigain* (three twenties – 60) and *pedwar ugain* (four twenties – 80). There are also examples of other written forms: *pum ugain* (five twenties – 100) and *chweugain* (six twenties – 120), and there is also some evidence of the oral use of multiples of twenty as far as *ugain ugain* (twenty twenties – 400) in some parts of Wales a hundred years ago and more. These are interesting, but rare, examples. So, yes, Welsh does have a well defined method of counting in twenties but it falls short of being a comprehensive system by comparison with that developed by the Mayans.

It would be nigh on impossible to argue that the traditional Welsh vigesimal system meets the demands of the modern digital age, but, as we shall see in the next chapter, the process of modernisation was no mean achievement.

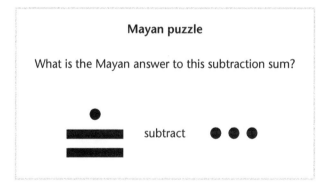

Mayan puzzle

What is the Mayan answer to this subtraction sum?

7 WHAT DO YOU RECKON?

THAT NIGHT in 1820 was unusual, even by the standards of the regulars at the Red Lion who had witnessed many hard-fought contests over the years. But tonight, two of the villagers, Hwfa and Guto, were going head-to-head.

Hwfa was conservative by instinct, always ready to defend the old traditions, including the old way of counting, or reckoning as Hwfa preferred to call it. Guto was a moderniser who gave short shrift to ancestral methods: it was high time that the Welsh adopted an easier way of counting. He had been heartened by recent correspondence in the Welsh press praising the virtues of the 'new' method, particularly the letter written under the pen name of 'Llewelyn o Abertawe' (Llewelyn from Swansea) that hit the nail on the head: 'The English can count slowly to a thousand before any Welshman, no matter how much heat his mouth generates by his efforts, can reach a mere four hundred' (my translation).

Fired up by this public encouragement, Guto was ready to challenge Hwfa's conservative stance. His confidence was slightly dented when the big night arrived as he sensed that not all of the regulars at the Red Lion were on his side. Only about half of them, if that, appeared to support his revolutionary argument while the other half – the behemoths of their time, in Guto's view – were hoping that this brazen upstart would get his comeuppance.

The rules of the contest are agreed: the first to reach 200 wins. Guto and Hwfa stand face to face and, on the referee's signal, they start counting in unison *un, dau, tri,...* (one, two, three,...). After reaching ten (*deg*) Hwfa continues, using the traditional method of counting: *un ar ddeg, deuddeg, tri ar ddeg, pedwar ar ddeg, pymtheg,... ugain, un ar hugain,...* (one on ten, two on ten, three on ten, four on ten, fifteen,... twenty, one on twenty,...). Guto follows the new method: *un deg un, un deg dau, un deg tri, un deg pedwar, un deg pump,... dau ddeg, dau ddeg un,...* (one ten one, one ten two, one ten three, one ten four, one ten five,... two tens, two tens one,...). The crowd bays its encouragement to both contestants to the rhythmic sound of the counting – particularly those who have placed bets on the outcome.

By the time they reach a hundred there is little in it. Beads of sweat begin to appear on both their foreheads and Guto is surprised at Hwfa's strong grasp of numbers such as *pedwar ar bymtheg a thrigain* (four on fifteen on three twenties) for 79, compared to Guto's apparently simpler *saith deg naw* (seven tens nine). But the strain is beginning to tell on Hwfa and, despite the encouragement of his supporters, the effort of having to concentrate hard on each number is taking its toll – perhaps he shouldn't have had that second pint after all. As they pass 150, Guto's steady rhythm gets the upper hand and he reaches the 200 mark while Hwfa remains bogged down in the 170s. Guto's supporters give a roar of applause that reverberates around the neighbouring streets and the few Englishmen at the inn are astonished at the fervour of the battle.

Something of the kind represents the first recorded efforts to reform the Welsh numbering system to respond to the new industrial age of the nineteenth century. Notwithstanding the changing demands of the workplace, that wasn't in fact the main force for change. Rather, the levers for reform were more heavenly than earthly and were found in the needs of chapels and churches as they conducted their Sunday services up and down the land. Congregational singing was becoming increasingly popular and there was a boom in the publication of hymnals. One of the early publishers was the Reverend David Jones, a Nonconformist minister in Holywell, Flintshire. His first hymnal, published in 1820, contained 500 hymns and, for ease of reference, they needed to be numbered systematically. But the idea of putting a number such as 492 above a hymn was too revolutionary for David Jones, who opted instead for CCCCXCII – Roman numerals in their full glory.

Imagine the confusion in the congregation: a hymnal that shows the Roman numeral CCCCXCII and the minister announcing hymn number *'pedwar cant deuddeg a phedwar ugain'* (four hundred twelve and four twenties). There was no option but to press for a change in the system that allowed the minister to announce hymn number *'pedwar cant naw deg dau'* (four hundred nine tens two) so that the congregation could easily understand that 492 was meant. Ministers came under pressure to adopt the new system and to persuade the members of their congregations to use the new numbers widely in their everyday lives as well as during services. That broader aim had little success but there was a clear change in the way in which hymn numbers would be read and used in Welsh.

The pattern has stood the test of time in Welsh chapels and churches: the norm is to use the decimal system to announce hymns but, extraordinarily, to continue to use the traditional vigesimal system to announce a reading from the scripture. For example: *'Darllenir o'r ddeunawfed bennod ar hugain o Lyfr y proffwyd Jeremeia ac yna fe ganwn rif yr emyn tri deg ac wyth.'* ('We shall read from the two-ninth chapter after twenty in the Book of the prophet Jeremiah and then we shall sing hymn number three tens and eight.') And there we have it: the number 38 being expressed in words in two different ways in the same sentence.

Normal practice is to refer to chapters using ordinal terms – first, second, third and so on – but to use the more usual cardinal numbers when referring to hymns – one, two, three and so on. The chapters of a book have a set order as they reveal their story in a sequential fashion. By contrast, the hymns in a hymnal do not follow a set order that determines why, say, hymn 36 should come before hymn 37. Hence, ordinal numbers are used for chapters and cardinal numbers for hymns.

Beyond announcing hymns in chapels and churches there was little impact in nineteenth-century Wales on the traditional ways of expressing numbers in Welsh. While there was a vigorous correspondence in the press there was no institutional backing that could promote a wider dissemination of a simpler method of counting. By the time that schools for the children of ordinary people were established later in the nineteenth century, the Welsh mindset had been influenced by the publication of a report commissioned by the government on the state of education in Wales. Commonly branded as *Brad y Llyfrau Gleision* (The Treason of the Blue Books), this report, published in 1847, had a profound and lasting influence on attitudes to the use of Welsh in education: Welsh needed to be sidelined in favour of English if children were to be able to make progress in the wider world. English became increasingly recognised as the lingua franca of trade and industry, the language of education in schoolrooms and college lecture theatres, and the main language to be used on the street, in shops and in the home. This mindset continued virtually unchallenged until the end of the Second World War. To what extent this can be attributed directly to the Blue Books, rather than that the Blue Books merely captured the aspirations of the Welsh themselves, remains a contentious issue in Welsh historiography. For generations of pupils, the outcome was that, outside religious services, counting and calculation were things that one did in someone else's language.

In today's Wales, the ways in which children and adults use Welsh when naming numbers can depend on many factors, including the age of the person, the part of Wales in which they live, and whether or not they've had the advantage of a Welsh-medium education. I undertook a small survey of a representative sample of individuals in Wales, asking each person to read out sentences in Welsh that contained numbers written in figures rather than in words. In administering the survey I tried to make sure that individuals gave responses that corresponded to what they would say naturally rather than what they thought the researcher wanted them to say.

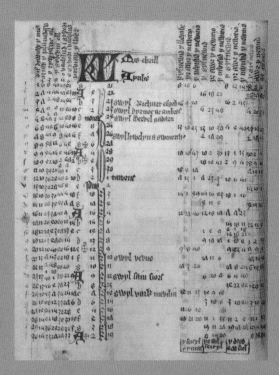

Plate 1. *Gutun Owain's use of Hindu–Arabic numerals in 1488/9 to list the feast days for the month of April (mis Ebrill).* (National Library of Wales. NLW MS 3026C, p. 16)

Plate 2. Magical Symbols, *an interpretation by the Pembrokeshire artist Anne Gregson, showing the use of Recorde's '=' sign in iconic equations. The painting can be seen at the Tenby Museum and Art Gallery.*

Plate 3. *(right)*
A parchment
representing the
date 18 May 1976
on the Mayan
calendar.

Plate 4. *(below) The*
pupils of the Rawson
school, 1880. (Library
and Archives Service,
Bangor University)

Plate 5. *A star pattern on a quilt made by Sara Lewis, Aberdare, in 1875.* (From the Jen Jones collection. Photograph by Roger Clive-Powell)

Plate 6. *Portrait of William Jones (1674–1749) by William Hogarth (1740).* (National Portrait Gallery)

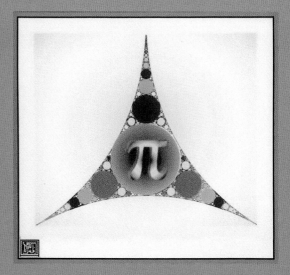

Plate 7. *Image by Jan Abas – a tribute to William Jones*

Plate 8. *Equally Puzzled.* (Courtesy of Claudia Williams and Martin Tinney Gallery)

As an example, one of the sentences was '*Rwyf yn byw yn rhif 20, Stryd Bangor*' ('I live at number 20, Bangor Street'). In this particular case, the number '20' was read in one of three ways: *ugain* (twenty) – using the traditional vigesimal system; *dau ddeg* (two tens) – using the modern decimal system; or reverting to the English 'twenty'. The patterns that emerged overall from this limited study included:

- Older people tended to use English number words for the more 'complex' numbers, such as 17 and 25. But this group was also the most likely to use the traditional vigesimal words in other contexts. (Most of the individuals in this group had received their early education through the medium of English and had been taught to do their sums in English. Welsh remained their natural everyday language and they would use the traditional Welsh number words for the simpler numbers, turning automatically to English when they wanted to refer to a more complex number.)

- Members of the younger generation tended to use decimal Welsh number words – *un deg saith* (one ten seven) for 17, *un deg wyth* (one ten eight) for 18 and so on – rather than the traditional vigesimal equivalents, and avoided using English words. (Members of this group had received their early education through the medium of Welsh, and had used Welsh to do their sums. They were, therefore, more comfortable using the decimal forms of numbers.)

- A handful of the traditional Welsh number words were used commonly across generations, particularly *un ar ddeg* (one on ten) for 11, *deuddeg* (two on ten) for 12, *pymtheg* (five on ten) for 15, and *deunaw* (two nines) for 18. (These particular words are heard daily on the radio and television, in conversations on the street, at home and at school. They are universally accepted and widely used.)

The survey also included examples of using numbers in the context of calculation, such as '*Y rhif sydd un yn fwy na 17 yw 18*' ('The number that is one more than 17 is 18'). In these cases no respondents used the traditional forms of the numbers 17 and 18, and they divided between those, typically of the younger generation, who naturally used the modern decimal forms, and others, typically of the older generation, who reverted to the English number words.

After the Second World War, education and the media, particularly radio and television, have been the strongest influences on the way in which individuals say their numbers when they talk in Welsh. With the sustained growth of Welsh-medium education over that period it is by now very common for young children to learn their mathematics in Welsh, and a decision was made consciously by educators during the late 1940s that the decimal method of saying numbers would be used in schools because this was the easiest way by a long chalk of developing children's understanding of number. Welsh radio and television have also developed ways of referring to numbers that incorporate a mix of both the modern decimal system and the traditional vigesimal system.

Recent research has confirmed that children in Wales who learn how to count in Welsh using the decimal system have a better early understanding of how numbers work than those in Wales who learn how to count in English. The logically simple structure of the decimal system of counting in Welsh gives them a clear advantage – just as it does in the languages of the countries of the Far East such as China, Japan and South Korea.

Nevertheless, the practice of using the decimal system of counting in schools at the expense of the traditional vigesimal system has not been without its critics in Wales. One of the most vociferous of those critics was Iorwerth Peate (1901–82): poet, scholar and first director of St Fagans National History Museum, located on the outskirts of Cardiff. For Iorwerth Peate the thought of encouraging children to say *tri deg saith* (three tens seven) for 37 rather than *dau ar bymtheg ar hugain* (two on fifteen on twenty) was anathema and an example of 'talking English in Welsh'. Iorwerth Peate laid the blame firmly and squarely on schools as being 'the chief culprits in the process of abandoning the traditional Welsh way of counting in twenties' (my translation). Heated letters were published in the Welsh press during the 1970s giving both sides of the argument. One of those letters challenged Iorwerth Peate to give the Welsh for the English 'forty-eight thousand' (48,000). He immediately and dismissively responded with '*wyth mil a deugain*', literally 'eight thousand and two twenties'. His challenger then retorted, tongue in cheek, by asking what was the Welsh for 'eight thousand and forty' (8,040). The sting in the question arises because *wyth mil a deugain* is ambiguous: it can mean both 'eight thousand and forty' and 'eight and forty thousand' and is therefore both 8,040 and 48,000. Iorwerth Peate chose to ignore the challenge and the debate fizzled out. Nevertheless, some language experts continued to argue that the new decimal words were 'alien to Welsh'.

The modern decimal system has now become firmly established but many Welsh speakers continue to use both methods of referring to numbers, much in the same way that metrication has not completely banished the everyday use of imperial measures and the associated vocabulary.

Examples that use both systems, even within the same sentence, are commonplace. For example, the pre-recorded announcements at railway stations in Wales are often made in both English and Welsh. The Welsh version of one such announcement at Cardiff railway station was heard to be '*Y trên nesaf i adael platfform tri fydd y dau ddeg un munud wedi un ar ddeg o'r gloch i Gaergybi*', closely followed by its English equivalent, 'The next train to leave platform three will be the twenty-one minutes past eleven train for Holyhead'. The curious feature of the Welsh announcement is that the form *dau ddeg un* (two tens one) for twenty-one follows the decimal system of counting whereas the form *un ar ddeg* (one on ten) is the traditional way of expressing eleven. Here, then, we had both methods within the same sentence, but no one missed their train as a result.

In her description of the Welsh international footballer John Charles, who played for Leeds and Juventus, Mererid Hopwood writes: *'Enillodd John Charles ei gap cyntaf yn ddeunaw oed. Aeth ymlaen i gipio tri deg saith arall.'* Literally: 'John Charles won his first cap when he was two nines old. He went on to win a further three tens seven caps.' In this example Mererid Hopwood uses the traditional method when referring to age (*deunaw*) and the decimal system to state the number of caps won (*tri deg saith*). In contexts that involve time – someone's age, the time on a clock – there is a strong element of putting things in order ('the march of time') and, just as in the case of referring to chapters of the Bible, Welsh appears to be more comfortable using the traditional system of counting when using ordinal numbers. People tend to read the time as *'pum munud ar hugain wedi un ar ddeg'* ('five on twenty minutes after one on ten') rather than *'dau ddeg pump munud wedi un deg un'* ('two tens five minutes after one ten one') and are more likely to ask, *'Pryd wyt ti'n cael dy barti pen-blwydd yn ddeunaw?'* ('When are you having your two nines birthday party?') rather than *'Pryd wyt ti'n cael dy barti pen-blwydd yn un deg wyth?'* ('When are you having your one ten eight birthday party?').

But times they are a-changing and the situation is open to many external influences, including that of the digital age. If, for example, a digital clock shows the time as '11:19' it would be common these days to hear that read using the decimal system as *'un deg un, un deg naw'* ('one ten one, one ten nine') although some would be more comfortable saying *'mae hi bron yn ugain munud wedi un ar ddeg'* ('it's almost twenty minutes past one on ten') and yet others, who may be less comfortable using Welsh for their numbers, might mix both languages, saying *'mae'n eleven nineteen'* ('it's eleven nineteen') or *'mae hi bron yn twenty past eleven'* ('it's almost twenty past eleven'). A variety of combinations can be heard, as in the example of the train announcement, but the underlying pattern is that Welsh speakers tend to use the decimal system when referring to the number of objects (how many?) – the cardinal number – and the traditional method when referring to the order of things (which position?) – the ordinal number.

The media have responded to and assisted these changes in many ways. Under the strong leadership of Thomas Davies during the early 1960s, the Sports Department of the BBC in Wales adopted a policy directing announcers, when reporting scores in, say, rugby or cricket, to use the traditional system of counting for numbers up to thirty and to change to the decimal system for higher numbers. On a Saturday afternoon it was possible to hear scores such as '*Caerdydd wyth ar hugain, Pontypridd tri deg chwech*' ('Cardiff eight on twenty, Pontypridd three tens six') – both methods in a single announcement. This practice is no longer followed and revised BBC Wales guidelines, covering all programmes rather than only sports programmes, encourage broadcasters to make 'sensible' use of both methods.

The boundary between the use of the modern decimal system on the one hand and the traditional vigesimal system on the other continues to shift. It can, for example, be influenced in practice by the social context of a conversation. When our son, Huw, was in his early teens he was on the phone one Saturday morning with a school friend arranging a time and place to meet. In the middle of the conversation, Huw turned to me to ask, '*Dad, fedri di fynd â mi i Fethesda erbyn hanner awr wedi un ar ddeg?*' ('Dad, can you take me to Bethesda by half past one on ten?') Having got my agreement to this request, Huw turned back to confirm the arrangement with his friend, '*Ia, grêt! Wela i di ar y sgwâr am half 'leven.*' ('Yes, great! I'll see you on the square at half 'leven.') Huw had transferred from the somewhat more formal Welsh with his father to re-establish his street 'cred' with his schoolmate. Our choice of number words is both complex and fascinating.

On the field of the National Eisteddfod held in Bala in the summer of 2009 I had a brief conversation with Elfyn Pritchard, chair of the Eisteddfod's literature committee that year. Elfyn was as pleased as Punch with the response to one of the main poetry competitions, rejoicing that the Eisteddfod had received as many as '*deugain a phedair o bryddestau*' ('two twenties and four odes'), and going on to say that it was unusual for the Eisteddfod to receive more than '*tri deg o gynigion*' ('three tens entries'). Here again, both methods of counting are being used in the same sentence, but this time to refer to the same thing, the number of poems submitted for a competition. What factors, I wondered, had influenced Elfyn's choice of words? 'Well', he responded, 'it just depends how I feel.' By now the rules have slackened and you make your own choice.

Well, not entirely! The ordinal numbers in Welsh – those for first, second, third and so on – remain troublesome. Take the case of how you refer to the nineteenth century in Welsh. We have two options, one based on using the modern decimal system and the other based on using the traditional vigesimal system. Using the decimal system we get *yr un deg nawfed ganrif* (the one ten ninth century), which is perfectly logical and unambiguous. Using the traditional method, on the other hand, yields *y bedwaredd ganrif ar bymtheg* (the fourth century on fifteen), which is more unwieldy and yet more 'poetic'. Traditionalists raise their eyebrows at the use of the modern method in such contexts – it just doesn't sound right, they argue. In order to sidestep the issue, publishers often use the abbreviated '*19g.*' ('19c.') in print, leaving it to the reader to decide how to express it in words. For those who had their early mathematics education in English the tendency is to use English in such situations. If you listen carefully to a conversation in Welsh you can sometimes pick out the English words for numbers, when the speaker is uncertain regarding the correct form to use or, indeed, when the speaker prefers not to show off his or her understanding of some of the convolutions of the traditional forms.

Proponents of the modern decimal system to express ordinal numbers as well as the more familiar cardinal numbers continue to press their case, arguing that the traditional system is far too complex, particularly for large numbers. For example, the ordinal number ninety-ninth (99th) can easily be expressed using the modern method as *naw deg nawfed* (nine tens ninth), whereas the traditional form would require *pedwerydd ar bymtheg a phedwar ugain* (fourth on fifteen and four twenties). A radio sports commentator was trying to ask one of the competitors in the annual Snowdon race to confirm that he was competing in his thirty-first race. The modern form for thirty-first would have yielded the simple *tri deg unfed* (three tens first) but, in an attempt to be linguistically correct, the poor soul tied himself up in knots trying to use the more traditional form, leaving his listeners to second-guess what exactly he was trying to say. The debate continues and we have no option but to rerun the counting contest first staged in 1820, but this time using ordinal rather than cardinal numbers.

We can imagine the competitors facing up to each other on the stage at the National Eisteddfod: Mod on the left raising the banner on behalf of the decimal system and Trad on the right arguing the case of the vigesimal system. They have both been practising in preparation for the battle and the tension in the pavilion is palpable. At the signal for the off given by the Archdruid, resplendent in her fine robes, they start in unison: *cyntaf, ail, trydydd,...* (first, second, third,...). After reaching *degfed* (tenth) together, Trad continues on his vigesimal route: *unfed ar ddeg, deuddegfed, trydydd ar ddeg, pedwerydd ar ddeg, pymthegfed,... ugeinfed, unfed ar hugain,...* (first on ten, two tenth, third on ten, fourth on ten, five tenth,... twentieth, first on twenty,...). Mod follows the decimal system: *un deg unfed, un deg eilfed, un deg trydydd, un deg pedwerydd, un deg pumed,... dau ddegfed, dau ddeg unfed,...* (one ten first, one ten second, one ten third, one ten fourth, one ten fifth,... two tenth, two tens first,...). The audience roars its support to the beat of the rhythmic counting. But the dilemma faced by many in the audience is that they can't decide whom to support. I know where I've put my money.

Vigesimal puzzle

Pre-decimalisation in 1971, why was it common to refer in Welsh to a ten-shilling note as a *'papur chweugain'* (a 'six twenties note')?

By contrast with English-speakers, Welsh-speakers make decisions about themselves and their listeners, consciously or otherwise, every time they use number words. Does that add to the richness of the experience of dealing with numbers in Welsh or does it add neuroticism to their numeracy?

8 PRAIRIE POWER

REAT NEWS, a baby boy! Mother and child are well. A cause for celebration and an opportunity to put an announcement in Welsh in the local paper:

On 4 May a baby boy, Ioan Gwyn, was born to Tina and Huw Gwyn in Trevelin, weighing 4,050 kg. (my translation)

Four thousand and fifty kilograms? Impossible! The paper must have made a printing error – four thousand and fifty kilograms is about four tons, the weight of a middle-aged elephant. The truth slowly dawns. This is an announcement in *Y Drafod*, a Welsh-language paper that is circulated to Welsh speakers in Patagonia, in a region of Argentina where an adventurous group of Welsh men, women and children established a community in 1865. The announcement in *Y Drafod* had been placed by descendants of those original settlers. The metric system has long been in use in Argentina where, in common with many other countries, a comma is used when writing sums of money or noting weights and other measures – for example, $1,30 and 5,7 kg where, in Britain, a full stop is used: £1.30 and 5.7 kg. In Argentina, 4,050 kg means 4 kilograms and 50 grams. Ioan is a healthy, bouncing baby boy.

Misunderstandings across cultural barriers are common, even when those cultures have shared roots, as George Bernard Shaw remarked about the English (*sic*) and the Americans: 'Two peoples separated by a common language'. Something similar may be said about the Welsh in Wales and those of Welsh ancestry in Patagonia, particularly regarding their number vocabulary. It would be all too easy to assume that those who emigrated from Wales, having formed their own community in Patagonia, would have tried to preserve their language patterns and other traditions. In the world of mathematics (and education in general) they in fact made innovative advances, leaving their fellow countrymen in Wales far behind. By now Welsh speakers in Wales are wholly conversant with the decimal system of counting in Welsh and only use the older vigesimal system, as we have seen earlier, in very particular contexts such as when referring to age. Welsh speakers in Patagonia, by contrast, threw out the old system and adopted the 'modern' decimal system within a short period of settling in their adopted homeland.

In 1940, when Gwilym Thomas was a young sailor, he joined the crew of a ship that sailed from Wales to Argentina. Taking advantage of some shore leave after reaching Buenos Aires he decided that he would try to contact descendants of the original Welsh settlement, reasoning that some of them would by then have moved to live in the country's capital. Gwilym had no direct contacts that he could get in touch with, but he had sufficient confidence to look in the Buenos Aires telephone directory for the first Jones he could find. He had a stroke of luck – the number he tried was answered in Welsh and the Jones family immediately invited him to visit them in their flat in the Buenos Aires suburbs. But they were in for a surprise when Gwilym arrived. They had been expecting a middle-aged man, rather than a comparative youngster. The misunderstanding caused much hilarity and leg-pulling, and Gwilym was given a typically warm Welsh welcome. But what had been the cause of the misunderstanding in the first place? When Gwilym had made his first contact by telephone, he had explained that he was a nineteen-year-old sailor and, quite naturally, had used the traditional Welsh way of referring to his age, *pedair ar bymtheg* (four on fifteen). The Jones family had no idea what this number was – they were only familiar with the modern decimal system of counting in Welsh – and their best guess was that they had heard *pedwar ar bum deg* (four on five tens), equivalent to 4 + 50 or 54. It was therefore not surprising that the family were expecting a middle-aged gentleman and were surprised – pleasantly surprised, one can guess – when they opened the door to a young nineteen-year-old.

The explanation for the misunderstanding is easily understood: the method used to count by the Welsh settlers and their descendants in Patagonia is different from that traditionally used back home in Wales. After the first settlers had landed at Porth Madryn in 1865 and had begun to build up their community they were free to decide on their own way of life, on their methods of religious worship, on how they wanted their children educated, and on how they would develop and run their businesses and organise their work. This was to be a Welsh-language-based settlement in accord with the vision of the founding fathers, Michael D. Jones and Lewis Jones. That principle would inform all aspects of the settlers' everyday lives.

In 1868, R. J. Berwyn (1836–1917) established one of the first schools for the children of the settlers and he was eager to put the founders' vision into practice. Welsh was the natural language of the school and all the lessons, including those on mathematics, were taught in Welsh and only in Welsh. While their monoglot Welsh cousins back home were doing their sums in English and reciting their 'twice one, two; twice two, four...', the children of the settlers were learning '*dau un, dau; dau dau pedwar...*'. On his visit from Wales to the settlement in 1882, Michael D. Jones was astonished to hear children reciting their multiplication tables in Welsh, an experience that was entirely new to him. Doing their sums in a language that was foreign to them was the experience of children in Wales; doing their sums in their first and natural language was the experience of the young Patagonians.

The decision to work with numbers in Welsh meant not only that schools taught sums in Welsh but that all buying and selling transactions in local businesses were completed and recorded in Welsh. The settlers established a limited liability cooperative trading company (*Cwmni Masnachol y Camwy Cyf.*) to serve their needs. The annual accounts of that company dating from the late 1800s are quite extraordinary. They show that the settlers had developed their own Welsh vocabulary to cover all aspects of accountancy and they used these new terms with confidence. During the same period, the use of Welsh in similar contexts in Wales would have been highly unusual. There, English, and English alone, was the language of business and trade.

But how could all this be achieved – teaching children to do their sums in Welsh and using only Welsh in business and trade – if they were stuck with the clumsy traditional method of naming numbers in Welsh? Would children in Patagonia have to calculate 36 + 74 by saying the equivalent of 'one on fifteen on twenty add four on ten on three twenties'? That would be absurd. The way ahead was to adopt a decimal system of counting. Using that system the sum would now become the equivalent in Welsh of 'three tens six add seven tens four', one that is so much easier to juggle. As a result, a quiet revolution began in faraway Patagonia. In 1878 R. J. Berwyn published a small book for teachers setting out a number of guidelines regarding what should be taught to children. This was the first book to be published in Welsh in Patagonia and R. J. Berwyn used the opportunity to recommend a decimal method of counting. That became the basis for the pupils' number work: avoid the use of traditional words associated with the vigesimal system of counting, and keep it simple by using only the decimal system.

I had been aware of these differences before visiting Patagonia in 2002, and was keen to see and hear for myself the use made of Welsh when modern-day Welsh-speaking Argentinians deal with numbers. There have been so many influences on Welsh in Patagonia in the intervening years that it's not easy to draw sound conclusions. The period between about 1868 and 1906 were the golden years for Welsh in Patagonia. By 1906 the Argentine government had decreed that Spanish should be the main language of education. As a consequence, Spanish is the language of choice of Patagonians today, including many of those who are able to speak both Welsh and Spanish. Nevertheless, Welsh-speaking Patagonians readily understand Welsh numbers using the decimal system: *dau ddeg pump* (two tens five) for 25, for example. They can tell you the year in Welsh, saying, for example, '*mil naw cant wyth deg pump*' ('thousand nine hundred eight tens five') for 1985, when many Wales-based Welsh speakers would automatically turn to the English 'nineteen eighty-five' and others to the abbreviated *un naw wyth pump* (one nine eight five).

Welsh-speaking Patagonians are, however, less sure of the traditional forms of counting in Welsh. Some have difficulty understanding *deuddeg* (twelve) and *pymtheg* (fifteen), while they are perfectly happy with the decimal *un deg dau* (one ten two) and *un deg pump* (one ten five) for the same numbers. As one Patagonian explained to me, 'The word *deuddeg* sounds very much like *dau ddeg* [two tens = twenty] and the word *pymtheg* sounds too much like *pum deg* [five tens = fifty].' That, of course, explains Gwilym Thomas's problem. By today many Patagonians use the word *ugain* (twenty), particularly when referring to age or to a date, and some have encountered it when reading. But the big bugbear is the word *deunaw* (two nines), the traditional way of saying eighteen. I heard virtually no one using this word in Patagonia. Curious to know more, I asked some of the older Patagonians if they were familiar with the meaning of *deunaw*. The response was overwhelmingly that the word was completely unknown to them.

Home and school aside, another important language influence on generations of Patagonians has been the chapel. There, children would hear the language patterns of the traditional Welsh Bible. They would hear the word *deuddeg* (twelve) often enough: the twelve tribes of Israel; Jacob's twelve sons; Jesus's twelve apostles. They would also hear references to numbers that include the word *ugain* (twenty): the forty days of the flood (*deugain* = two twenties); Christ fasted for forty days and forty nights. But where would they hear the word *deunaw* (eighteen)? Eighteen occurs only rarely in the Bible and not once in the New Testament. Would this explain why the word *deunaw* has disappeared virtually completely from the vocabulary of Welsh-speaking Patagonians? For them, eighteen is invariably *un deg wyth* (one ten eight).

Gaiman is an important town in the history of the Welsh settlement in Patagonia. Located in the Chubut Valley, the town has a small museum, the Museo Histórico Regional. Discussions began in 1960 to establish the museum, financial support was secured from the Argentine government, and Tegai Roberts (1927–2014), a descendant of Michael D. Jones, was appointed as its curator, a post that she occupied with distinction for over fifty years. The museum is housed in Gaiman's old railway station and its survival and development is a testimony to Tegai Roberts's hard work. The museum's collection includes unique documents, books and photographs that together trace the Welsh settlement's history. Among the documents is a remarkable school copybook kept by Caradog Jones.

Caradog was a pupil in his teens at R. J. Berwyn's school in Rawson in the 1880s (Plate 4). His teacher at the school was T. G. Prichard and he followed R. J. Berwyn's guidelines enthusiastically. Caradog learned to count using the new decimal system and to perform all kinds of calculations in Welsh. He also learned how to work with fractions, decimals and percentages, all through the medium of Welsh. But Caradog's lessons went much further than that. He also learned how to square numbers: for example, he could show that the square of 13 (13 × 13) is 169. More remarkably, he also learned how to work the process backwards. Starting with the number 106,929 Caradog found that 327 is the number that, when squared, makes 106,929. In modern terminology, Caradog had learned how to calculate square roots, and he did it all in Welsh.

To cap it all, Caradog used his knowledge about numbers to solve problems in geometry. In particular he was able to apply Pythagoras' theorem to right-angled triangles. The Greeks were familiar with the theorem from around 500 BC and it's possible that it was also known to the Babylonians and the Chinese a thousand years earlier. In our time the theorem has been central to the mathematical education of generations of pupils, although no one in Wales in 1880 would have been able to use the theorem through the medium of Welsh. But here, locked away in a cupboard in the Gaiman museum, was evidence that Caradog Jones in 1880 was completely confident using the theorem in his mother tongue.

In his copybook, Caradog uses a range of newly-minted Welsh terms to describe the problem that he's trying to solve about a right-angled triangle, whose hypotenuse is of length 52 yards and its upright of length 20 yards, and goes on to apply Pythagoras' theorem to show that the length of the base is 48 yards. This is an example of a problem that is now solved daily by secondary-school pupils, and the Welsh terms, although by now different from those used by Caradog, are familiar and well entrenched. But this was 1880 and no one in Wales at that time – no one at all – would have been doing anything remotely like this in Welsh.

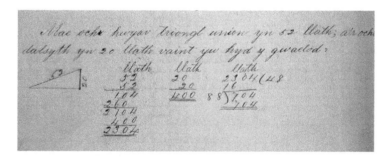

Figure 5. *Part of Caradog Jones's work in 1880 as he applies Pythagoras' theorem in the Rawson school.* (Museo Histórico Regional, Gaiman)

R. J. Berwyn showed that it was perfectly possible and practical to use Welsh as the medium of instruction across all subjects. In particular, he showed that calculations could be performed easily and naturally in Welsh and that the language was no barrier at all to teach mathematics. The whole scheme lasted until about the end of the nineteenth century when the Argentine government's policies weakened the role of Welsh within the education system. In the first half of the twentieth century some schools in Wales began to experiment by using Welsh to teach mathematics but it wasn't until the 1950s that schools began to adopt such a policy more widely, supported by appropriate resources. The pioneering work of R. J. Berwyn and his contemporaries was finally being emulated this side of the Atlantic.

One and one make two in Patagonia, as in Wales and everywhere else. But the experience of counting and of mathematics more generally has varied considerably across the generations in both countries. In Caradog Jones's time, mathematics was not alien to the culture of the Welsh settlers and children experienced it in their mother tongue. They did not have to suffer it being taught to them through the language of a different people – by contrast with the experience of children in Wales in the same period.

Prairie puzzle

Two gauchos, Arturo and Benito, buy horses and sheep at the local fair.
Arturo buys three horses and five sheep.
Benito buys five horses and three sheep and pays 4,800 pesos more than Arturo.

How many more pesos is the price of a horse than a sheep?

PUTTING DOWN DIGITAL ROOTS

WHAT DRIVES our obsession for pattern – in numbers, in words, in music, in art, in science? Why, for example, should we be intrigued by the pattern in one of the best-known English palindromes: A man, a plan, a canal – Panama? The best-known palindrome in Welsh is *Lladd dafad ddall* (literally, 'Kill a blind sheep') and a quick search of the web confirms that a fascination with palindromes is common to many languages.

Od nad wyf i fyw dan do (literally, 'I am strangely destined not to live under a roof', a sentiment ascribed to a wandering tramp) is a particularly interesting Welsh palindrome. Not only does it satisfy the requirement that it should read the same backwards as forwards but it also obeys the rules of *cynghanedd*, a formalised system of strict meter peculiar to Welsh that underscores the composition of traditional Welsh poetry. Those rules have developed over many centuries and such poetry is as popular and vibrant today as it was in the days of Dafydd ap Gwilym (*fl.* 1340–70), one of the earliest masters of the art. In this example, typical of the style, the line has seven syllables and the first three consonants 'd n d' at the beginning of the line are repeated at the end. The full set of rules that govern *cynghanedd* allows a rich mix of repeated patterns of consonants and internal rhymes that reflect the natural rhythm of the Welsh language.

Some words are themselves palindromes – personal names such as Bob and Nan, and common words such as bib, civic, deified, madam. But it's far more difficult to devise a palindromic phrase or sentence, particularly one that makes much sense. Try it yourself to see how difficult it is: the search is itself a fascinating exercise.

The basic idea behind palindromes can easily be extended. For example, the numbers 282 and 1,661 and 253,676,352 are all palindromic numbers. In music, composers, including Mozart and Haydn, were inspired to include palindromic tunes in their works.

Palindrome puzzle

2002 was a palindromic year.
What's the next palindromic year?

On a digital clock, what's the first palindromic
time after midnight?

Our instinctive drive to understand the world about us leads us naturally to look for patterns, in the belief, rightly or wrongly, that there are laws that determine the nature and form of that world and that those laws are not the whimsical musings of an eccentric superbeing set on confusing us with a bewildering set of random events. The challenge for scientists is to find patterns in what is observed. That elementary particles, for example, have an inbuilt symmetry is a basic assumption of physicists who explore the secrets of the atom, a human adventure that continues at centres like CERN, deep underground at Geneva.

Artists use patterns created by colour, shapes and forms to fire the imagination of the viewer. For poets, the patterns of rhythm and rhyme add a depth to the words. Equally, spotting patterns in numbers and shapes is central to the mathematician's ways of thinking. Simplicity and 'beauty' are always the main objectives: 'There is no permanent place in the world for ugly mathematics', wrote the mathematician G. H. Hardy.

This instinct to use pattern is central to traditional Welsh culture, both through the Welsh language and otherwise. As well as within Welsh poetry, as shown by *cynghanedd*, patterns enrich

Figure 6. *A Celtic knot pattern on a cross near Llandeilo, Carmarthenshire.*

traditional forms of folk singing, particularly in *cerdd dant*, which intertwines a melody sung by the singer with a different melody on the harp accompaniment – a perfect musical double helix. Celtic knot patterns adorn our ancient stone crosses and Welsh quilts (see Plate 5) exploit the potential of colourful patterns.

Being aware of and sensitive to patterns in mathematics also greatly helps in its learning. Spotting a pattern assists the *understanding why* as well as being a source of pleasure and enjoyment in itself. Think, for example, of the 9-times table – one nine is nine (1 × 9 = 9), two nines are eighteen (2 × 9 = 18), three nines are twenty-seven (3 × 9 = 27), and so on. Learning those bare facts alone misses the opportunity to see patterns in the table.

Here are the numbers in the 9-times table, as far as ten nines:

9, 18, 27, 36, 45, 54, 63, 72, 81, 90

What patterns can you see in these numbers? There are many possibilities, including:

- The line of numbers looks very much like a palindrome.
- The numbers are in pairs: 45 and 54; 36 and 63; 27 and 72 and so on.
- The units digit decreases by one every time: 9, 8, 7, 6, 5, 4, 3, 2, 1, 0.
- The sum of the digits in each number is always 9.

This last observation is particularly interesting because it is true of *every number* in the line. The instinct of the mathematician is to look for general patterns, ones that are true every time, without exception. What is suggested by this pattern of numbers is that the digits of the numbers in the 9-times table *always* add to give 9. This claim is certainly true of the numbers in the table between 9 and 90. For example, the sum of the digits of 45 is 4 + 5, which is 9, and the sum of the digits of 90 is 9 + 0, which is also 9. In order to make it easier to record this we shall use an arrow rather than writing 'add the digits of the number' each time. Using the arrow, we have $45 \longrightarrow 9$, and $90 \longrightarrow 9$. This is another example of the power of a symbol to represent a mathematical idea, similar to Recorde's use of '=' to replace 'is equalle to' (chapter 4).

But what happens if we extend the numbers in the table further than 90? After all, the 9-times table doesn't suddenly end at 10 × 9. Here are the next ten numbers in the table:

99, 108, 117, 126, 135, 144, 153, 162, 171, 180

What are the sums of the digits of each of these numbers, noticing that we now have to add up to three digits each time? Using the sign again, we can write $99 \longrightarrow 18$, $108 \longrightarrow 9$, and so on. The full list is:

$$18, 9, 9, 9, 9, 9, 9, 9, 9, 9$$

And, yes, the answers are still 9, in almost every case. The only exception is the 18 at the start, the sum of the digits of 99. But the sum of the digits of 18 is also 9 (1 + 8 = 9). We can write this as 99 $\longrightarrow$ 18 $\longrightarrow$ 9. There is something interesting going on here but we need to tidy it up before we can arrive at a pattern that works *every time*.

Let's go about things like this. Adding the digits of a number creates a new number and, if we continue the process by adding the digits of that number as well, we eventually arrive at a single-digit number. For example, if we start with the number 7,869 we generate the following steps:

$$7,869 \longrightarrow 30 \longrightarrow 3$$

And if we start with the number 7,868 we arrive at:

$$7,868 \longrightarrow 29 \longrightarrow 11 \longrightarrow 2$$

In the first example it takes two steps before we reach a single-digit number; three steps are needed in the second example.

In order to simplify the notation further and avoid having to say 'add the digits of the number repeatedly until we reach a single-digit number' we shall use the phrase 'digital root of the number'. We can then say that the digital root of 36 is 9, the digital root of 72 is 9, the digital root of 99 is 9, and we can use the same method to find the digital root of any number. For example, the digital root of 42 is 6, the digital root of 133 is 7, and the digital root of 2,132 is 8. Sometimes two or more steps are needed to find the digital root, as we saw above in the case of the numbers 7,869 and 7,868 – the digital root of 7,869 is 3, and the digital root of 7,868 is 2.

We can now go back to the numbers in the extended 9-times table:

9, 18, 27, 36, 45, 54, 63, 72, 81, 90, 99,
108, 117, 126, 135, 144, 153, 162, 171, 180,...

The final dots are used to show that the numbers in this list carry on indefinitely. Having introduced the idea of the 'digital root of a number' we can now venture to suggest something that may be true for every number in the table, that:

> the digital root of *every number* in the 9-times table is 9 and, conversely, if the digital root of a number isn't 9 then that number isn't in the 9-times table

Using some elementary algebra it is possible to prove that this statement is true – but you must take my word for that. The pattern holds true, not just for *some* of the numbers in the table, but for *every number*. That's a powerful statement to make and is an example of the power of mathematics itself.

To illustrate the power of this statement about the 9-times table, here are some examples that show how this general result can be applied in specific cases:

The digital root of 423 is 9, since 423 $\longrightarrow$ 9.
It follows that 423 is in the 9-times table.

What's the digital root of the number 3,799?
Adding the digits produces the pattern:
3,799 $\longrightarrow$ 28 $\longrightarrow$ 10 $\longrightarrow$ 1.
The digital root of 3,799 is 1.
It follows that 3,799 is *not* in the 9-times table.

What's the digital root of the number 3,878,514?
Adding the digits produces the pattern:
3,878,514 $\longrightarrow$ 36 $\longrightarrow$ 9.
The digital root of 3,878,514 is 9.
It follows that 3,878,514 is in the 9-times table.

We've shown that 3,878,514 is in the 9-times table without having to do any multiplications or divisions.

A word of warning: these patterns don't work for *every* times table. For example, the digital root of the number 17 is 8, but 17 isn't in the 8-times table. Beware!

On an occasion when I was giving a talk to a society at a local chapel I used this hymn board as a prop, one that is similar to the hymn boards used in Sunday services at scores of chapels and churches up and down the land.

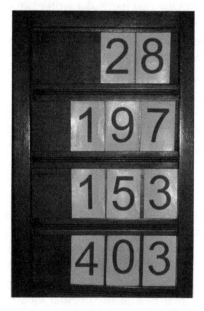

Figure 7. *'We shall sing hymn number…'*

Without having any clear idea of the response I would get, I began by asking if any of those present were in the habit of looking at the numbers facing them every Sunday in order to do anything with them mentally in any way – possibly before the service began or, even, during a particularly lengthy sermon. Much to my surprise, the members of the audience were eager to share some of their hitherto secret thoughts:

I look to see if all of the numbers between 0 and 9 are there. (On this particular board the only number missing is 6.)

I try to see how quickly I can add up the four numbers in my head. (The total of the numbers on this board is 781.)

A very unexpected response came from Morien Phillips, then a retired primary-school head teacher and a well known actor on the Welsh stage. He explained that it was his custom to experiment by finding 'the digital root' of numbers on the board – although he didn't use that particular term or any other technical jargon to describe it. On this particular board he would start by imagining lining up the four numbers to make one very large number:

$$28,197,153,403$$

This is the number 28 billion, 197 million, 153 thousand, 4 hundred and 3. He would then calculate its digital root, as follows:

$$28,197,153,403 \longrightarrow 43 \longrightarrow 7$$

And so the digital root of the number at today's service is 7.

But he didn't stop there. Morien Phillips went on to explain that he would find the digital roots of each hymn individually, going on to add the four roots and to calculate the digital root of the answer. To his constant amazement he always got the same answer as the digital root of the original long number. On this board, these would have been his steps:

28	$\longrightarrow 10 \longrightarrow 1$
197	$\longrightarrow 17 \longrightarrow 8$
153	$\longrightarrow 9$
403	$\longrightarrow 7$
	$1 + 8 + 9 + 7 = 25$
	The digital root of 25 is 7

Finally, as if that wasn't enough, he added the numbers of the four hymns to get the total of 781 and then found its digital root, giving 781 $\longrightarrow$ 16 $\longrightarrow$ 7. The answer 7 appears yet again. Morien Phillips repeated this experiment Sunday after Sunday, discovering that the pattern never failed, and marvelling at the simplicity of it all. With a little bit of algebra it is possible to prove that this is no coincidence: the 'trick' works *every time.*

On the wall of the stairwell in the library at Bangor University there's a poem, in both English and Welsh, in which the poet Gwyn Thomas praises the disciplines that are traditionally taught at universities. In his reference to mathematics, Gwyn Thomas refers to 'the elegance and fascination of numbers'. Universities are not the only settings in which we can enjoy mathematical patterns that create this 'elegance'. They are patterns that can be appreciated and enjoyed by all, as Morien Phillips experienced.

The digital roots of numbers also had very practical applications. As they closed their account books at the end of each day, back in the days before computers, bank clerks had to make sure that they hadn't made any errors in their calculations. One of the methods that they used to check their sums was to apply what corresponded to Morien Phillips's method. If the digital root of the total at the bottom of every page didn't correspond to the sum of the digital roots of each individual entry, then they knew that an error had crept in and that they needed to identify it. At one level, using Morien Phillips's method is a pleasant pastime to while away a few quiet moments in church, and it is all the more surprising to find that it had a practical use. Work done by mathematicians today may be used tomorrow in ways that we are completely unable to predict. For example, the security of plastic bank cards that we use daily at ATMs or to pay for items in shops and cafes and on the internet depends on the properties of prime numbers discovered by mathematicians hundreds of years ago, long before banks existed.

What about the other tables? Would similar patterns emerge if we calculate the digital roots of the numbers in the 8-times table, say, or the 6-times table? And that sets us off in pursuit of other patterns and we begin to act like real mathematicians. The essence of mathematics is to experiment by looking for patterns and it is that search that can enrich the subject for schoolchildren, for college students and for other inquisitive individuals like Morien Phillips.

Sometime during my first two years at secondary school I happened to stumble across an 'interesting' number pattern while doing my maths homework. I had noticed that adding the numbers 1, 2 and 3 gives the same answer as multiplying them – adding them gives $1 + 2 + 3$, which is 6; and multiplying 1, 2 and 3 gives $1 \times 2 \times 3$, which is also 6. Wow! I was completely mesmerised by this 'discovery' and the following morning was anxious to share it with my maths teacher. 'Did you know, sir', I asked him on the corridor between lessons, 'that if you add one, two and three, you get the same answer as when you multiply them?' 'Yes', he replied, 'I know that', and walked away. The teacher lost a golden opportunity to take advantage of the natural enthusiasm of a young pupil. I was completely deflated and the disappointment was so much keener because of my admiration of him as a teacher. But in all fairness, he probably had other things on his mind that morning as he rushed from one lesson to the next. In retrospect, the teacher would also have been conditioned by the restrictions of a mathematics syllabus that was largely confined to the learning of a seemingly endless list of techniques and the memorisation of formulae. There was no time, in reality, to 'stand and stare' at the magic of patterns in number and shape. As a school subject at that time, mathematics could be lifeless and uninspiring.

By today, noticing patterns in mathematics is a natural part of children's experience at primary school. Opportunities also occur at secondary school, both within GCSE courses and at A level, and in college and university programmes. It is unlikely that such was the experience enjoyed by Morien Phillips during his own schooldays, but he was still ready to venture and to marvel at the patterns revealed to him every Sunday. The challenge for teachers and lecturers is to present mathematics as an exciting subject that can capture the imagination of schoolchildren and college students alike. One way of doing so is to encourage them to look for the wealth of patterns that hold the subject together. Like the painter and the poet, the mathematician is, above all else, a maker of patterns.

1, 2, 3 puzzle

The sum of the three numbers 1, 2 and 3 is equal to their product:

$$1 + 2 + 3 = 1 \times 2 \times 3$$

Can you find two numbers whose sum is equal to their product?

Can you find four numbers whose sum is equal to their product?

AREAS OF (MIS)UNDERSTANDING

APART FROM golden sands, what else links the Mediterranean island of Sicily and the isle of Anglesey off the north-west coast of Wales? More precisely, what is the mathematical link between the two islands? To concentrate your mind, imagine yourself strolling along one of Anglesey's beaches enjoying a cornet topped with a scoop of local homemade ice cream.

Over two thousand years ago, the Greek mathematician Archimedes (*c*.287–212 BC) discovered formulae to calculate some of the properties of curved shapes like cones and spheres (in three dimensions) and circles (in two dimensions). Born in the town of Syracuse on the island of Sicily, Archimedes excelled in mathematics and science and is considered to have been one of the very best mathematicians ever.

Think of the challenge of finding a way to calculate the area of a circle. Finding a formula for the area of a rectangle – 'length × breadth' – is child's play compared with the challenge of calculating the area of a shape with a curved perimeter. How can we possibly begin to set about it? Archimedes had a flash of inspiration that enabled him to calculate both the area of a circle and its perimeter (the length of its circumference). Applying his insight to three-dimensional shapes he then found ways to calculate the surface areas of shapes like cones and spheres as well as how much space they occupy – their volumes. His methods have enabled succeeding generations to measure the size of such diverse objects as earthly ball bearings and cannon balls, as well as the size of the moon and other heavenly bodies. As you enjoy your cornet you can even use Archimedes' formulae to calculate how much ice cream fills the cone and how much is in the hemisphere that tops it.

It is remarkable that all of these formulae make use of a very special number, whose value is slightly greater than 3. Archimedes was familiar with this number and knew its approximate value. By today we know very much more about it and that its value is 3.141592... where the dots show that the decimal continues forever.

π puzzle

14 March is Pi Day, celebrated on the same day every year.

Why on that particular date?

Almost two thousand years after Archimedes had developed his formulae, William Jones (1674–1749), a self-taught mathematician from Anglesey, suggested using a special symbol to represent

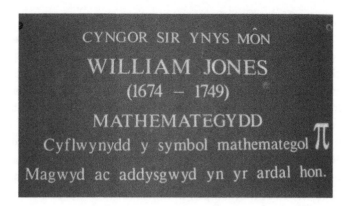

Figure 8. *A slate plaque, written in Welsh, on the wall of the primary school in Llanfechell, Anglesey. Erected by Anglesey County Council, it celebrates that the mathematical symbol π was introduced by William Jones, who grew up and received his early education in the area.*

Archimedes' number. This symbol, the Greek letter π (pi), appears for the first time in 1706 in a book on mathematics written by William Jones.

Archimedes' creativity in Sicily over two thousand years ago, combined with the precision of Anglesey's William Jones three hundred years ago, is celebrated daily in our secondary schools when pupils are first introduced to two formulae about circles:

the circumference of a circle = the circle's diameter × π
the area of a circle = the circle's radius × the circle's radius × π

You may remember these formulae from your school days as:

the circumference of a circle = πD *or* the circumference of a circle = 2πr
and
the area of a circle = πr^2

(In these formulae, 'D' represents the length of the diameter of the circle and 'r' represents the length of its radius. D = 2r, since the length of the diameter is double the length of the radius.)

Does this ring a bell?

Higher up the school, some pupils will also be introduced to Archimedes' formulae for curved three-dimensional shapes, including the sphere and cone. The feature that is common to these formulae is that they all use the remarkable number π. Primary-school children also use a variety of ways, including practical methods, to calculate areas and volumes but they are unlikely to be using Archimedes' formulae. In their early years children need to enjoy a range of practical experiences to enable them to get a feel for and develop an understanding of what is being measured, for example, that the area of a two-dimensional shape is a measure of its *size*. Being able to remember a formula is not the most important aim. Using the terminology of our first chapter, using a formula is *knowing how* but *understanding why* is much more of a challenge.

William Jones (see Plate 6) is one of Anglesey's most famous sons. He was born on a smallholding close to the village of Capel Coch in the parish of Llanfihangel Tre'r Beirdd, north of the county town of Llangefni in the middle of the island. When William was still a small child the family moved a few miles further north to the village of Llanbabo. He attended the charity school at nearby Llanfechell, where his early mathematical skills were drawn to the attention of the local squire and landowner, who arranged for William to go to London, where he worked in a merchant's counting house. He was shortly sent on a commercial voyage to the West Indies, an experience that began his interest in navigation.

When he reached the age of twenty, William Jones was appointed to a post aboard a man-of-war to give lessons in mathematics and navigation to the crew. On the back of that experience, he published his first book in 1702 on the mathematics of navigation as a practical guide for sailing. On his return to Britain he began to teach mathematics in London, holding classes in coffee shops for a small fee. Shortly afterwards, in 1706, he published his seminal work, *Synopsis palmariorum matheseos*, a book written in English, despite the Latin title, which may be roughly translated as 'A summary of achievements in mathematics'. The symbol π appears in this book for the first time to denote the ratio of a circle's circumference to its diameter. In Greek, π is the first letter of the word for periphery (περιφέρεια) and π is also the first letter of the word for perimeter (περίμετρος). It is thought that one or the other influenced his choice of this particular symbol. William Jones suspected that the decimal 3.141592... never ends and that it cannot be expressed precisely. 'The exact proportion between the diameter and the circumference can never be expressed in numbers', he wrote. That was why he recognised that it needed its own symbol to represent it. The symbol π was popularised in 1737 by the influential Swiss mathematician Leonhard Euler (1707–83), but it wasn't until as late as 1934 that the symbol was adopted universally. By now, π is instantly recognised by school pupils worldwide, but few know that its history can be traced back to a small village in the heart of Anglesey.

William Jones became friendly with Sir Thomas Parker, later the Earl of Macclesfield, and tutored the young George Parker, who was to become the second Earl. He later lived at the family

There are various other ways of finding the *Lengths*, or *Areas* of particular *Curve Lines*, or *Planes*, which may very much facilitate the Practice ; as for Instance, in the *Circle*, the Diameter is to Circumference as 1 to

$$\frac{16}{5} - \frac{4}{239} - \frac{1}{3}\overline{\frac{16}{5^3} - \frac{4}{239^3}} + \frac{1}{5}\overline{\frac{16}{5^5} - \frac{4}{239^5}} -, \&c. =$$

3.14159, &c. $= \pi$. This *Series* (among others for the same purpose, and drawn from the same Principle) I receiv'd from the Excellent Analyst, and my much Esteem'd Friend Mr. *John Machin* ; and by means there-

Figure 9. *An example of William Jones's use of π to denote the ratio of the diameter of a circle to its circumference, from* Synopsis palmariorum matheseos, *p. 263.* (Library and Archives Service, Bangor University)

home, Shirburn Castle, near Oxford, where he developed close links with the family. Through his numerous connections William Jones amassed at Shirburn an incomparable library of books on science and mathematics. He also maintained links with Wales, particularly through the Morrises of Anglesey, a family of literary brothers renowned for their cultural influences and activities who, although a generation younger than William, came from the same part of Anglesey and had strong London-based connections.

In the wake of publishing his *Synopsis*, William Jones was noticed by two of Britain's foremost mathematicians: Edmund Halley (who had a comet named after him) and Sir Isaac Newton. He was elected a Fellow of the Royal Society in 1711 and was vice president of the society during part of Sir Isaac Newton's presidency. William Jones became an important and influential member of the scientific establishment. He also copied, edited and published many of Newton's manuscripts. In 1712 he was appointed a member of a committee established by the Royal Society to determine whether the Englishman Isaac Newton or the German Gottfried Wilhelm Leibnitz should be accorded the accolade of having invented the calculus – one of the jewels in the crown of contemporary mathematics. Not surprisingly, considering the circumstances, the committee adjudged in favour of Newton.

William Jones married twice. One of the children of his second marriage was born barely three years before William died aged 74. Also named William Jones – a source of much subsequent confusion – the son became Sir William Jones, appointed as a Supreme Court judge in India and an expert in the languages of the subcontinent. Sir William established links between Latin, Greek and Sanskrit, thus establishing the concept of 'Indo-European languages' that remains a cornerstone of modern linguistic theory. He was once introduced to the king of France as one who knew every language apart from his own – Welsh! It is highly likely, given his place of birth, that William Jones, the father, would have been fluent in both Welsh and English but, having lost his father when he was three, Sir William would not have had the opportunity to learn his father's first language.

William Jones the elder died in London in 1749 and was buried in the churchyard of St Paul's, Covent Garden. In his will he bequeathed his library of roughly 15,000 books together with some 50,000 manuscript pages, many in Newton's hand, to the third Earl of Macclesfield. Some 350 of these books and manuscripts were written in Welsh, and this portion of the original library was safeguarded in about 1900 to form the Shirburn Collection at the National Library of Wales in Aberystwyth. A hundred years later, in 2001, that part of William Jones's collection that comprised papers and notebooks belonging to Sir Isaac Newton was sold to the library of the University of Cambridge for over £6 million, a sum partly raised by public subscription. The bulk of the rest of the library was sold in a series of auctions at Sotheby's, London, in 2004 and 2005, raising many more millions: a copy of the astronomer Johann Kepler's *Harmonices mundi* raised close to £100,000 and Newton's classic *Principia mathematica* a further £60,000. William Jones's own book, *Synopsis palmariorum matheseos*, was a bargain at a mere £8,000. In one of Newton's books, edited by William Jones, and given as a gift by Jones to the Macclesfield family, there was a single loose sheet in Newton's own handwriting. This sheet alone raised £90,000. The Macclesfield estate benefited greatly from the sale, but this priceless collection has now been dispersed to libraries and private collectors across the globe. Some mystery remains regarding the fate of William Jones's personal papers. It appears that the Macclesfield family had been reluctant to release them and there is the suggestion of a scandal that the family has sought to conceal. Those papers would surely throw further light on William Jones, on his relationship with the earls of Macclesfield, and on his remarkable life-journey from a cottage in Anglesey to be a member of the mathematics establishment, and one of its shining stars.

William Jones's work, particularly his insight into the importance of the number π, continues to delight mathematicians today. Dr Jan Abas (1936–2009), a lecturer at Bangor University's Department of Mathematics, became entranced by William Jones and devoted himself to raising awareness of his work and his talent. The image in Plate 7 is a tribute to William Jones by Jan Abas and is rich with symbolism: π is placed at the centre of a pattern that has rotational symmetry of order 3, the circles are measured using π, and the sequence of ever-decreasing circles highlights the infinite nature of π.

As part of my work as a mathematics adviser I had an opportunity to visit a primary school in the middle of Anglesey, not far from William Jones's birthplace. In a class of ten-year-olds I sat with a group of pupils who were working hard on 'area', the focus of their mathematics lesson that morning. Betsan was working her way diligently through a page of problems, each problem requiring her to calculate the area of a rectangle: sometimes presented as a lawn to be sown, sometimes as a field to be ploughed and sometimes as a wall to be painted. But the same principle underpinned each question, that of applying the formula:

area = length × breadth

Betsan's first calculations were relatively straightforward, such as calculating the area of a lawn measuring 3 metres long and 2 metres wide. As Betsan worked her way from one problem to the next the calculations became progressively more difficult and included challenges such as 17 × 23 and some that involved decimals, such as 6.5 × 2.4, but the method remained the same: multiply both numbers together.

A number of plastic shapes were scattered on Betsan's table – a collection of squares, triangles and circles. Talking with Betsan about her work I picked up one of the circles and placed it on the cover of a book that was also on the table, before asking Betsan, 'Which has the larger area, the book cover or the circle?' For me, the answer was obvious because the book cover was clearly larger than the circle. But Betsan was perplexed, before hesitantly suggesting that 'the circle hasn't got *any* area'. That was not, of course, the answer that I had been expecting, and I began to think that perhaps Betsan hadn't heard the question correctly, so I rephrased it slightly, 'Take a look at the book cover and at the circle that's lying on the cover. Which of the two has the greater area, the cover or the circle?' Betsan again hesitated before answering: 'No, the circle hasn't got *any* area.' Rather than correcting her – the natural reaction – I probed a bit deeper, 'Why do you say that?' 'Well', began Betsan, more confidently by now, 'the circle hasn't got any length or breadth.'

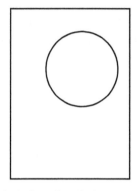

Which shape has the larger area?

The cat was out of the bag! For Betsan, 'area' was synonymous with the formula 'length × breadth'; she had only limited understanding that the concept of 'area' is a measure of a shape's size. Rather, for her, the formula was the concept. If a shape had neither 'length' nor 'breadth' it naturally followed that you couldn't calculate its area – the concept was meaningless in such a context. Betsan reasoned well but, starting from the wrong place, she was bound to lose her way. She had spent the lesson doing multiplication sums and had multiplied small numbers, larger numbers and decimals. But she had not had to confront the main purpose of the lesson – that of coming to terms with area as a concept. Superficially she had mastered the topic, answering every question correctly, the teacher's line of ticks testifying to that success, but her response to deeper questions revealed an underlying lack of understanding of what she was doing. The trick of *knowing how*, provided the shapes were rectangles, had been mastered by Betsan but her grasp of *understanding why* was far less secure.

'Area' is a difficult concept to master; many of us are content with *knowing how*, hoping that we've chosen the correct formula. By the end of her time at primary school, Betsan, in common with many other children, interpreted 'area' as a formula. On reaching secondary school, would that be her expectation of mathematics more generally – that it is just a matter of learning a set of techniques to get the answers, rather than understanding the concepts that underpin those techniques? Would that also be her experience of using formulae involving π, the symbol first used by the local boy, William Jones? *Knowing how* but not *understanding why*?

CRACKING THE CODE

MY MEMORY of that day remains crystal clear, every detail having left a lasting impression. It started as a fairly ordinary Tuesday morning in early October 1984. I had arranged to visit Ysgol Gymraeg Morswyn, a Welsh-medium primary school on the outskirts of Holyhead, to enjoy the company of the children and to discuss mathematics with the teachers. The head teacher, Islwyn Williams, welcomed me to the school and I began by visiting the infants class and sitting around a table with a group of five-year-olds.

Seeing some small cubes scattered on the table, I was prompted to play a simple game with the children. After choosing five of the cubes and getting the group to count them carefully I placed my hand over some of the cubes leaving just one cube in view. That suggested the question: 'I wonder how many cubes are hiding under my hand?' After much thought and guesswork, there was no shortage of possible answers from the children: two, three, four, and none at all (after all, none could be seen, could there?) Finally, I lifted my hand and we all counted how many there were – four, of course. We had a lot of fun with other numbers of cubes and the children themselves took it in turns to hide cubes and to ask the others how many they were hiding.

It soon became clear that the game posed no challenge at all to one of the children, although Gareth hadn't wanted to spoil their fun by answering on behalf of the others. He was sufficiently sensitive, although only five years old, to know when to answer and when to hold back. As we tidied up, the class teacher whispered to me that Gareth was 'rather good' in maths and that perhaps I would like to have a few words alone with him. Those few words developed into a quarter of an hour's conversation. I began by adapting and extending my original question by asking, 'Imagine, Gareth, that I've got a hundred sweets and I give you one of them. How many do I have left?' That question was also far too easy for Gareth and in no time at all I was asking him, 'What if I had a *million* sweets and give you one? How many would I have left then?'

After an instant's thought Gareth gave me his answer but it wasn't the one I'd expected. Before reading further, pause for a minute to decide what answer you would have given to the question. It's far from being easy and many people – schoolchildren and adults alike – would struggle with it. The question is a good test of our understanding of numbers – of *understanding why* – and of how numbers are organised in the decimal system:

ten	10
hundred	100
thousand	1,000
ten thousand	10,000
hundred thousand	100,000
million	1,000,000
and so on	

The big challenge for typical seven-year-old children is to understand that a number like 37 is 3 tens and 7 ones (or units), and that this is different from 73, which is 7 tens and 3 ones. The technical term to describe this is 'place value'. As children progress through primary school they need to be able to understand place value in numbers over a hundred and then over a thousand. By the end of primary school some children, but not everyone by a long way, will also understand place value in numbers over a million. In an experiment with ten-year-olds, children were shown a picture of a car's milometer showing the number of miles that the car had travelled:

0	2	6	9	9

The children were then asked to write down what numbers would be on the milometer after the car had travelled one more mile. Fewer than half the children (48 per cent) gave the correct answer. In other words, more than half the children showed that they were having difficulties understanding 'place value' with numbers over a thousand. Very, very few ten-year-old children could have answered the question that I'd asked Gareth: what number is one less than a million?

Are you now ready to read on?

The answer in figures to the question that I'd asked Gareth is 999,999 – the comma showing where the thousands start. The convention of adding a comma is not universal and other variants are used in different cultures to help the reader to read large numbers. In Britain, the convention is to group the digits in threes and to use commas to separate the groups. For example, we can see immediately that £36,683 corresponds to a total of thirty-six thousand, six hundred and eighty-three pounds. For those who are familiar with this use of the comma, the natural way of reading 999,999 is in two parts:

> nine hundred and ninety-nine thousand,
> nine hundred and ninety-nine

But Gareth was only five years old. He hadn't come across the comma convention. That didn't stop him, however, but the form of his answer was different from what I'd expected. This is what he said (in the left-hand column), with virtually no hesitation:

'nine hundred thousand	900,000
ninety thousand	90,000
nine thousand	9,000
nine hundred	900
ninety	90
nine'	9

Gareth 'saw' the number clearly in his head and read it off, digit by digit. He *got* place value; *understanding why* came naturally to him.

By then we were both beginning to enjoy ourselves and Gareth had sensed that today's visitor to the school might be somewhat different. He fumbled in his pocket and pulled out a crumpled piece of paper on which were scribbled a lot of 1s (ones) and 0s (zeros). His next sentence floored me completely: 'I was reading last night about *binary numbers.'*

At one level, binary numbers can appear to be fairly elementary because they only use the digits '1' and '0'. But don't be deceived – understanding them properly is no mean achievement. We are familiar with counting in tens and using the numerals 0, 1, 2, 3, 4, 5, 6, 7, 8 and 9 – and that for the simple reason that we have ten fingers. Were humans to have only eight fingers – four on each hand, rather than five – then it is virtually certain that our number system would be based on counting in eights rather than in tens. (I had naïvely thought that this would be clear to everybody until I once got into difficulties with a group of college students who argued vehemently that counting in tens was the only possible way to count – as if our number ten had been divinely ordained as being the basis of our numbering. We had a long and lively discussion.)

In the binary system we count in twos rather than in tens, using only the numerals 0 and 1. As a result, binary numbers can look something like:

10111 (corresponding to the decimal number 23)

or

11000101 (corresponding to the decimal number 197)

and so on, explaining the scribbles on Gareth's piece of paper. Binary numbers are very useful, and indeed are essential, in computer science and are the basis for developments in that field. As a sequence of 0s and 1s they give the appearance of forming some kind of code and we are so familiar with numbers written in decimal notation that we have difficulty getting to grips with those same numbers written using a different number base. Fortunately we don't need to explore the detail of binary numbers in order to appreciate Gareth's achievement. It is quite remarkable that five-year-old Gareth had taught himself binary numbers. He had cracked the code. When I asked him what number was 'one less than a binary million' he wrote out the correct answer with no hesitation whatsoever.

Even more remarkable was that Gareth had not shown his piece of paper to anyone else in the school and certainly not to the infants teacher. After all, there was no expectation that she would be conversant with binary numbers, and Gareth was sufficiently sensitive to realise that and not to cause her any embarrassment. The common sense shown by Gareth was strikingly uncommon. As I visited other classes during the day Gareth followed me from room to room and joined in the discussions with children of all ages, understanding immediately what was going on. His level of *understanding why* was quite remarkable. It was a privilege to be in his company.

Binary puzzle

What number in binary is one less than the binary number 1000?

Your memory can play tricks on you. Some things, like my day with Gareth over thirty years ago, are imprinted in my memory as if they happened yesterday. Other memories remain hazy and distant. You can often persuade yourself that you've remembered something in minute detail whereas, in fact, your subconscious mind has filled in many of the gaps without telling you.

In the parable of the lost sheep I could have sworn that the New Testament story in Welsh refers to 'one less than a hundred sheep' left in the flock while the shepherd went to look for the one sheep that had wandered away. In Welsh there is a very convenient word that corresponds to the concept of 'less than' or 'falls short by'. That word is *namyn*. 'One less than a hundred sheep' would then be *cant namyn un dafad* (hundred less one sheep). The word is used in modern speech in a variety of contexts, such as the price of a garment being 'one penny short of ten pounds' (*decpunt namyn ceiniog* = ten pounds less a penny), or that a politician has spoken for 'a minute short of an hour' (*awr namyn munud* = hour less a minute). I first came across the short and convenient word *namyn* when listening as a child to the parable of the lost sheep; it fired my imagination as a snappy way of referring to a number that could otherwise be quite complex. In English the expression for the number that is one less than a hundred is itself quite snappy – ninety-nine. But in traditional Welsh, as we have seen in chapter 7, ninety-nine has to be expressed as 'four on fifteen on four twenties' – quite a mouthful. The Bible was first translated into Welsh in 1588 by Bishop William Morgan and the numbers in it are almost invariably expressed using the traditional vigesimal method. Indeed that first translation, which set the standard for written Welsh for almost four hundred years, itself raised the status of the traditional method of expressing numbers in Welsh. What a relief, then, that the translators could also often avoid the convoluted forms of numbers by using the handy *namyn*. In essence, the concept of *namyn* was the focus of my conversation with Gareth on my visit to Ysgol Gymraeg Morswyn, although I hadn't used that particular word. Gareth had calculated *cant namyn un* (hundred less one) and then *miliwn namyn un* (million less one) in a flash.

But my memory had failed me. The phrase *cant namyn un* does not in fact appear in the parable of the lost sheep as it is written in William Morgan's Bible. In Luke's Gospel, the number of sheep left in the flock is given as '*[n]amyn un pum ugain*', literally 'less one five twenties', showing the full force of the vigesimal system in which a hundred is replaced by 'five twenties'. By the time we reach more modern translations of the Bible published in the twentieth century the problem disappears as ninety-nine is now written as '*naw deg naw*' ('nine tens nine'), the decimal system replacing the vigesimal system and *namyn* made redundant, although, as we have seen, it still continues to be used in a variety of contexts both orally and in written form, and it retains a certain literary charm.

William Morgan's Bible includes a number of other uses of *namyn* in an effort to simplify the numbers, but they do not always strike a clear note. For example, in St John's Gospel an invalid lying by the side of the pool of Bethesda is said to have been ill for '*namyn dwy flynedd deugain*', meaning 'less two years two twenties', or thirty-eight years, which, in modern Welsh is *tri deg wyth* (three tens eight). In the book of the prophet Ezra we read that '*namyn tri pedwar ugain ŵyn*' ('less three four twenties sheep') have been sacrificed, which is seventy-seven sheep. One wonders how the Church fathers coped with such contortions, but they may have helped them to hone their mental-arithmetic skills.

The word *namyn* also features prominently in the Welsh version of the Anglican Church's *Book of Common Prayer*. The book lists the Church's thirty-nine articles of faith, referred to in Welsh as '*Namyn un deugain erthyglau crefydd*' or 'Less one two twenties articles of faith'. This title was first used in 1687 and is still in use today in the Church of Wales. In some parts of Wales *namyn* has been used in churches and chapels in a variety of other contexts. For example, there is anecdotal evidence that hymns were announced in chapels in some parts of Pembrokeshire during the 1960s using expressions such as *deunaw namyn un* (two nines less one) for 17, *deunaw namyn dau* (two nines less two) for 16 and *deunaw namyn tri* (two nines less three) for 15. These are remarkable remnants of a long tradition.

Some constructions in other languages show a similarity to the use of *namyn* in Welsh. For example, in Roman numerals 99 is written as IC, representing one less than hundred and the Latin word for 99 is *undecentum*, which, again, is one less than a hundred.

Namyn has a certain traditional charm and retains a place in modern Welsh, but its use in expressing large numbers is now rare and has been overtaken by the increasing use of the straightforward decimal system of writing and speaking numbers. In its heyday it certainly needed a nimbleness of mind to be able to decode the written and spoken word, similar to the depth of understanding shown by the five-year-old Gareth when he calculated one short of a million.

*In memory of Gareth Wyn Williams – see the
last page of this book.*

12 DOES MATHEMATICS HAVE A GENDER?

MAGINE THE SCENE. It's about six o'clock on a Saturday morning. My wife, Menna, and I are sound asleep. The alarm clock won't deprive us of a lie-in this morning. But it isn't to be. Our seven-year-old daughter, Llinos, comes to our room and asks her mother, 'Mam, what's four thirteens (4 × 13)?' Before reading further, pause to think what your response might have been to the question.

I've described this situation to many audiences, mostly at parents' evenings, and asked for their responses. Because the question had been directed at Menna, the women in the audience tend to be the first to respond. The most common response, given the early hour, is 'Go back to bed', but another frequent response is 'Go and ask your father'. The most common response from fathers in the audience is along the lines of 'Fifty-two (52), now go back to bed'. This pattern speaks volumes: mothers identify questions about numbers as being the responsibility of the father, and fathers need to verify their manhood by answering the question with a finality that closes further debate.

There is rarely any attempt to engage with the child, not surprisingly at that time of the morning, but, to be fair, some parents do offer more open responses, such as 'Why do you want to know?' or, less often, 'I wonder how we can work that out?' In this particular case, Menna managed to keep the conversation going. (I was listening intently from under the duvet, gradually realising that this encounter overlapped with my professional life and that reputations could be at stake.) After some discussion, Llinos saw that she could work out four thirteens (4 × 13) by looking at it as four tens (4 × 10) and four threes (4 × 3), or 40 add 12, giving a total of fifty-two (52). She happily went back to her room but came back within a few minutes to ask, 'What's four fourteens (4 × 14)?' By now Menna's response was more obvious, 'Can you remember what you did to work out 4 × 13? What about doing the same thing with 4 × 14?' 'Well', said Llinos on her third visit, '4 × 13 is too small and 4 × 14 is too big.'

By breakfast time we got to know what the questions were all about, and that Llinos had been trying to draw a colourful pattern on a sheet of paper marked out in fifty-four columns and wanted

Figure 10. *Llinos's present to her parents*

to divide the pattern into four equal sections. She was using her questions to try to find a *quarter* of fifty-four so that she could plan how many columns to include in each section. The whole point of the exercise suddenly became clear to her somewhat dim-witted parents, who could now see why 4 × 13 was too small and 4 × 14 too big.

In my experience, Welsh women are not alone in their willingness to admit openly that they have difficulties with mathematics. It appears to be socially acceptable to confess to being innumerate but completely unforgivable to admit to being illiterate. While this antipathy is not confined to women it appears that it is more likely to be expressed by women than by men. While I was buying a book on the history of mathematics at a local bookshop, the assistant, a young woman, remarked, 'I would never be able to read a book like that'. She didn't know what the book was about (apart from having seen the word 'mathematics' in the title) and yet she was happy to make her comment to a complete stranger.

It is also a common experience to hear radio and television presenters – both men and women, but particularly women – openly confessing to their fears about mathematics to their listeners and viewers. The context is often when a child is being interviewed about his or her favourite school subjects. The profile of mathematics within schools has greatly improved over the years and children are now much more likely to be positive about their mathematical experiences and to say that it is one of their favourite subjects. The instinctive, and devastating, reaction of some female presenters, of a different generation to the children being interviewed, is to say something like: 'Really? I was no good at all in maths when I was in school!' Male presenters appear to be less likely to fall into this trap, being conditioned, consciously or not, by some form of unspoken assumption linking manhood and mathematics.

A particular example may provide a clue to this behaviour. A very experienced male Welsh-medium radio broadcaster was listening on-air to a conversation between a teenager and an interviewer who had asked him the standard question about his school subjects. In his reply the teenager had said that he got a lot of enjoyment from playing chess and from mathematics. Commenting on the interview the broadcaster said that he never thought he would hear the words 'enjoyment' and 'mathematics' being used in the same sentence. He clearly thought that his off-the-cuff comment was amusing, but he may also have been aware, if only subconsciously, that his reputation as a broadcaster was not on the line. He might not have volunteered such a comment earlier in his career. Can you imagine that happening on local radio in, say, Beijing or Seoul or Singapore? Such behaviour would be completely baffling to an Asian audience.

Are men – particularly young men – less likely to be open about their attitudes to mathematics? Are there social pressures on men to behave as if they were more confident in the subject, as there are pressures on them to be comfortable doing science and technology? Men are the ones who can read maps, who know how the coffee machine works, who can cope with the technology of the barbecue, who can change a bulb and mend a fuse. Isn't that so?

A revealing example of an expression of this antipathy towards mathematics and science more generally is encapsulated in this *englyn* (an epigrammatic stanza that conforms to the rules of *cynghanedd*) composed by O. M. Lloyd (1910–80) when he was a 17-year-old school pupil:

> I hate 'that Mathematics' – I'm a dunce
> All my days in Physics.
> I cannot do Mechanics,
> Is my brain full of some bricks?

The quotation marks on 'that Mathematics' suggest that the poet is expressing his hatred here of his experience of mathematics and science *at school*, rather than his more general attitude to these subjects. O. M. Lloyd was a highly respected Nonconformist minister and a witty raconteur as well as being an accomplished poet. He only rarely wrote poems in English but it is significant that he chose to do so on this occasion. The choice reflects his experience of having been taught mathematics through the medium of English at school, as was the norm during his schooldays, and thereby associating mathematics with a culture that was alien to his natural daily experience.

The attitude of women to mathematics is a longstanding concern. Relatively few women, even today, reach top jobs in mathematics and science, whether in industry or in higher education. There are very few people who would argue that they do not have the ability to succeed in these fields, that their genes are somehow not fit for purpose. Nevertheless, the social expectations continue to impose glass ceilings and there are many organisations and campaigns devoted to seeking to address the imbalance. At Bangor University, for example, as at other universities, conferences on 'women in science' are held frequently to provide opportunities for schoolchildren to hear the experiences of women who have succeeded in these fields, such as Professor Siân Hope from Bangor, who specialises in computing. In America the Association for Women in Mathematics encourages women to study mathematics and to take up careers in the mathematical sciences. That association is as busy today as ever.

The tendency to label women as low achievers in mathematics and the expectation that men should be high achievers is deeply ingrained in our society. Relatively few women have achieved worldwide recognition as mathematicians, and the number of women from Wales who have received such recognition is very, very small.

The Department of Mathematics at the University of St Andrews in Scotland maintains a database of the world's most famous mathematicians, from Archimedes onwards. The list does not include people who are now living and it is striking that it only includes nine names from Wales, compared to a total of 309 from England, 190 from Scotland and 36 from Ireland. Can that be explained? It is also striking that only one of the nine from Wales is a woman – Mary Wynne Warner (1932–98).

Figure 10. *Mary Wynne Warner (1932–98).*

Mary Davies was born in Carmarthen, the elder of two daughters of Sydney and Esther Davies. Sydney Davies taught mathematics at Carmarthen Grammar School for Boys. When Mary was six years old, the family moved to Llandovery where her father had been appointed headmaster of the local grammar school. She excelled at her studies throughout her primary years and at her father's grammar school, and won the highest marks in Wales in her 16+ examinations. By then she had developed a liking for mathematics but she also wished to study physics in the sixth form. At that time the science facilities at Llandovery were sparse and the family decided to send Mary to Howell's, a residential school near Denbigh in north

Wales. Her father, meanwhile, had been appointed headmaster at Holywell Grammar School in Flintshire, so the whole family moved to north Wales. Some years later I became a pupil at Holywell Grammar School but was completely unaware of the mathematical prowess of the headmaster's elder daughter.

Mary Davies won a scholarship from Howell's to study mathematics at Somerville College, Oxford. After graduating, she married Gerald Warner, a fellow student who had read history. Following Gerald Warner's appointment as an ambassador in the diplomatic service, the family spent periods of time in a variety of countries, far and near – China, Burma, Poland, Switzerland and Malaysia. As a consequence of this nomadic lifestyle and the demands of bringing up a family, Mary Warner faced difficulties in developing her own mathematical interests and carving out her own career – a diplomat's wife's first priority was to support her husband. Nevertheless, Mary Warner succeeded in keeping up her work in mathematics and gained a doctorate (PhD) while the family lived in Burma – a remarkable achievement, particularly given that country's attitude towards the education of women. In Mary Warner's case, her husband had to use his full diplomatic skills to persuade the authorities to allow his wife to register for the degree. She specialised in a branch of modern algebra and made significant and original contributions to her field.

On eventual return to Britain, Mary Warner was appointed as a Reader in mathematics at City University, London, before being promoted to a Chair at that university – another remarkable achievement. She continued to publish prolifically up to her retirement in 1996 and thereafter. As a renowned academic she was often invited to speak at other universities. On one occasion in the early 1980s, she accepted an invitation to address a conference at the University of Bahrain. Having retired from the diplomatic service, her husband, now Sir Gerald Warner, decided to accompany her on the trip. In Sir Gerald's own recollections of the event:

> We were met at the plane by one of Mary's junior hosts, and taken to stay with one of my colleagues. Shortly after we had settled in I was called to the phone. It was our host, who told

me that he understood that Professor Warner was a woman. I confirmed that was the case. He said that this would be difficult, for she would be lecturing to male students, and meeting male colleagues. Would she mind if she was categorised as an 'honorary man'?

I naturally consulted Mary, who said she would have no objection. The visit passed off very pleasantly. Our final lunch was in the male quarters of our host's house. At the end of the meal I became an 'honorary woman', and was introduced to his wife and family in the female rooms.

Sir Gerald concludes that 'the occasion could only have involved a lady academic, and exemplified in a benign way the peculiarities of being a comparatively early bird as a lady mathematician'.

Mary Warner died suddenly while visiting friends in Spain in 1998. A tribute to her in *The Times* testifies that she didn't lose her sense of Welshness, despite having spent most of her life travelling the world. She was a plain speaker with a sharp sense of humour, but she sought to control her emotions in company to save the blushes of her husband, the ambassador. However, on one occasion she ran into some difficulties during a reception that they had arranged for diplomat guests during their period in Switzerland. They were at a hotel in Geneva that was famous for its *tartes à la crème*. Some of the guests were poking fun at Welsh poetry, much to Mary Warner's annoyance. She was unable to retaliate directly but decided to throw one of the hotel's cream concoctions at her defenceless husband. She had no rationale for doing so, of course, but her action certainly cut the conversation short.

For us, as pupils in her father's school in Holywell, a certain mystique surrounded the daughters of 'Syd'. It appeared to us that they kept their distance from Holywell and, although both Sydney and Esther Davies were members at our chapel, we heard little of their children. Mrs Walter Owen Jones was one of the oldest members of the chapel and when I won a place to study mathematics at Oxford she gave me a completely unexpected gift, a translation into English by Mary Warner of a book in French on algebra. At the time, I didn't realise that Mary Davies and Mary Warner were one and the same person, and it is clear in retrospect that Sydney Davies had given the book to Mrs Walter Owen Jones to transfer it to me. Sydney Davies was a complex character. He had never referred at the school to his daughter's work in mathematics, but had asked Mrs Walter Owen Jones, whose knowledge of algebra would have been minimal, to make sure that I received a copy of the book. One can only guess at what could possibly have prevented Sydney Davies from taking public pride in his daughter's success.

Mary Wynne Warner is the only Welsh woman to have received recognition from the University of St Andrews for her contribution to mathematics, the only Welsh woman included in its pantheon of the world's mathematical greats. And yet knowledge of her achievements is restricted to a relatively small group of like-minded mathematicians.

Very few girls at Holywell Grammar School specialised in science, including mathematics. In wasn't uncommon for secondary schools to bias the curriculum choices so as to encourage girls to make alternative choices. For example, pupils at Rhyl Grammar School during the 1950s had to choose between the subjects of scripture and mathematics with those choosing scripture, mainly girls, following a course in 'special arithmetic' that was devoid of any algebra or geometry. Many girls faced similar dilemmas during that period.

Mathematics was something for boys; 'softer' options appealed to girls – art, music, languages. This message was underscored openly and frequently. For example, in reviewing a mathematical puzzle book for children in 1932 the Yorkshire Post commented: 'Ought to be in every house where there are children. A regular gold mine for fathers, uncles, and schoolboys; and even sisters and mothers will be unable to resist some parts of it.' This assumption regarding the role of women is often contrary to the reality. For example, it was common in the farming industry for the farmer's wife to keep the farm accounts while her husband got on with the heavier work around the farm. The wife would oversee the bank accounts, and it was she who placed orders and who looked after the paper work. Such arrangements would often be part and parcel of family life more generally.

This became clearer to me over the course of a number of years in the case of two of my relatives, Gwen and her husband, Glyn. Both born in the 1910s, their educational opportunities had been limited. Gwen had worked as an office secretary before being employed as a bank clerk and then running a small village shop and post office. Gwen's number skills had been honed over the years and she continued to be able to perform mental calculations at speed throughout her long life. Glyn had worked as an accountant with the local gas board although he hadn't achieved the status of being a chartered accountant. As a young child I gradually came to realise that Gwen's number skills were better than Glyn's but that no one openly acknowledged that. It was a matter of pride for Glyn to establish his 'manhood' by showing that he was the one who understood mathematics and science. Gwen accepted her subservient role in these matters and avoided challenging or belittling her husband. Every Christmas night our extended family would congregate at our house and, as part of the evening's entertainment, play a number of party games. As the 'family mathematician' my role was to set the occasional number puzzle. Glyn felt that he should respond to the various challenges but it became clear that Gwen had the quicker mind. The challenge for me was to recognise Gwen's achievements while seeking to avoid embarrassing Glyn. I learned a great deal about the skills of diplomacy during those evenings.

On another occasion, after I had completed my doctoral thesis on the complex topic of the mathematics of the atom, Glyn was eager to see the thesis and asked to borrow it. When he returned it a few weeks later, he commented, 'Yes, very good!' He couldn't possibly have understood a single word of it (I have difficulty understanding it myself by now) but Glyn had felt that it fell to him to give it his blessing. Gwen was wise enough neither to comment nor to ask to see it.

Hafina Clwyd (1936–2011), author and frequent columnist in the press in Wales in both languages, was never slow to express her opinion and to do so firmly and clearly. The Welsh-language journal *Taliesin* had chosen mathematics as the theme of its Winter 2010 issue and I had contributed an article for the issue on the link between mathematics and culture. In her subsequent press column, Hafina Clwyd pulled no punches:

> The latest issue of *Taliesin* arrived. I am quite incapable of reading most of it. The theme is mathematics and it contains essays that make no sense at all and diagrams that make me feel lightheaded. In fact, since I was reading it in bed I had to get up to go to the bathroom to be sick because it brought back the whole nightmare of lessons on sums. Once again I was sitting dejectedly in the classroom staring at my work book and the red crosses against each answer. (my translation)

She claimed not to have understood a single word of my own article, even though it contained no mathematics as such. Hafina Clwyd has often referred in her autobiographies to her difficulties with 'sums' at school. Of her primary years at the village school in Gwyddelwern, she says: 'I would go home with my ears reddened by the close attention of the class teacher who had lost patience with a pupil who could recite poetry without hesitation but who couldn't for the life of her put two and two together to make eight.' (my translation)

Matters didn't improve when she reached the local grammar school where she began to feel physically ill at the thought of having to face maths lessons. Revealingly, she refers to pressures from her father who couldn't understand why his elder daughter was mathematically illiterate, and was bitterly disappointed that she was unable to gain entry to university because she hadn't chosen to study any science subjects and had failed her mathematics.

Hafina Clwyd wasn't alone; other women have also told me of their experiences of suffering physical symptoms at the prospect of doing mathematics. Some men may have had similar experiences but it appears that they are less likely to talk about them publicly.

Are such attitudes more common in Wales than elsewhere in the UK? Negative attitudes to mathematics, particularly among women, have long been the subject of discussion and research but it is difficult to make meaningful cross-border comparisons. However, a unique feature in Wales relates to the mix of languages used to teach mathematics in schools. Could this be a significant factor that has affected attitudes?

Teaching and learning mathematics in Welsh started fairly recently – essentially from the 1950s onwards. Before then, with few exceptions, only English was used to teach mathematics, even to children for whom Welsh was virtually their only language. Doing mathematics in a language that was foreign to them was their daily experience, and mathematics itself became a subject that was foreign to their natural culture. The seeds were sown many years ago that implanted in children the feeling that mathematics wasn't a natural part of their identity. Was this, ultimately, the root cause of the attitudes developed by Hafina Clwyd and others who could identify with patterns in poetry and song but who turned their backs on patterns in number and shape?

Patchwork quilt puzzle

An example of a puzzle published in 1933 that you would be unlikely to see today:

A number of young ladies made a patchwork quilt for the natives of Uganda. To celebrate completing the work, each girl kissed each other girl once, and the curate who was present kissed each of his sisters once. There were 72 kisses in all.

How many sisters does the curate have?

Claudia Williams is one of Wales's most respected artists, famous for her paintings and drawings that capture the domestic world of family gatherings, particularly of mothers and their children, often depicted as beach scenes.

Now living in Tenby, Claudia Williams was born in 1933 in Purley, Surrey, the great-granddaughter of a Cardiganshire sheep farmer. At Eothen School for Girls near Caterham, Surrey, she was particularly influenced, initially as a day pupil and later as a boarder, by the encouragement of her art teacher, Christine Walker, who had been a student of Graham Sutherland. Claudia Williams displayed an early natural talent and, even as a school pupil, won national prizes and scholarships including, at the age of 14, a prize in an art competition organised by the BBC's Welsh Children's Hour, an award that spurred her on to develop her talents further. At the early age of sixteen she attended Chelsea School of Art, where she won another scholarship. In 1953 she moved to north Wales where she married the artist Gwilym Prichard.

To celebrate the 500th anniversary of the birth of Robert Recorde in 2010 (see chapter 4), the Tenby Museum and Art Gallery invited artists to contribute to an exhibition of work that 'married art with mathematics and illustrated, using proportions, patterns, symmetry, perspective, geometry, form and line the importance of mathematics and, more specifically, Recorde's groundbreaking work'. One of the fifty or so works of art that were exhibited was an intriguing painting entitled *Equally Puzzled* by Claudia Williams (see Plate 8). In the catalogue for the exhibition, Claudia Williams comments:

> I was advised to give up the study of mathematics when I was twelve. Probably ill advised but I accepted the advice gladly at that point since it gave me extra time in the art room! So when I heard of this theme I must say that I felt rather uninspired given my experience at school. However after a while and knowing much more about Robert Recorde, plus the fact that he grew up in Tenby was enough to get me thinking about the subject. When I had done the painting, Gwilym came up with a quirky title for it.

A number of questions are prompted by the work. Assuming that the woman in the painting is the child's mother and that the child is her son, what has caused this state of being 'equally puzzled'? Could it be that both mother and son are puzzled in the sense of being bewildered by a page of mathematical symbols and diagrams? If only this homework, set by the son's maths teacher, could be dispensed with quickly and painlessly, they could both return to what they regard as more useful ways of spending their time.

A second, more positive, interpretation is that both mother and son are intrigued by the puzzle with which they are grappling and that the image captures a moment of joint reflection that leads to an exchange of ideas and a conversation that benefits them equally. Seen in this light, it is an example of what mathematics learning is all about – a joint endeavour and adventure – and one which emulates the example set in Recorde's dialogues between master and scholar almost five hundred years previously.

Which interpretation appeals to you more?

13 HOW TO MAKE MATHS REAL FOR ALL OF US

T SECONDARY SCHOOL, our chemistry master, Samuel Lynton Rees, was one of our favourite, if somewhat more eccentric, teachers. He had an infectious enthusiasm for his subject and wrote copious comments in red ink on our homeworks in his cramped handwriting. He had wide interests outside chemistry, particularly in music in which he had developed skills as a cellist. He also feigned interest in cricket and umpired school matches. His judgements from behind the stumps were not always reliable, but were invariably respected. 'Sammy' dubbed those of us having a particular interest in mathematics as *mathemagicians*, a term he used fondly and playfully. In retrospect, it's also a term that 'conjures up' a vision of a magic circle of wizards who have been initiated into the black art of mathematics, in which 'wizardry' sounds so much more positive than 'witchcraft'.

The image of mathematics as a mystical art linked with quasi-religious ideas has a strong historical pedigree. The Pythagorean school of ancient Greece that originated in the fifth century BC, inspired by Pythagoras (*c.*570–*c.*495 BC), was a religious movement at heart. It is thought that the Pythagoreans influenced the development of number mysticism and numerology. This irrational belief in divine and mystical relationships between numbers and daily events still resonates today and is exploited to great effect in popular culture by authors such as Dan Brown in his *The Da Vinci Code* best-seller.

A hundred years or so later than the Pythagoreans, the Greek philosopher Plato (427–347 BC) established his academy in Athens, the first institution of higher learning in the Western world. Above the entrance to that academy he inscribed the words 'Let no one ignorant of geometry enter here', a strap line that both recognises the importance of mathematics and suggests that access to its mysteries is restricted. Over the centuries from the classical period to the Renaissance, mathematical ideas were written about in western countries only in Latin and Greek and were inaccessible to ordinary people.

In the British Isles, Robert Recorde (chapter 4) was the first to challenge this convention by making mathematics available in English for 'the vnlearned sorte', those who had not had the advantages of a classical education. It is now regarded as a human right and a precondition for a modern society that its members should be both functionally numerate and have an awareness of the importance of mathematical ideas in the advance of science. As we have seen, however, we still suffer in the wake of Industrial Revolution thinking that rote learning the rules of basic sums was appropriate for the working-class masses while the delights of Euclidean geometry needed only to be made available to those enjoying a public-school education. Such thinking has also surfaced at other levels, for example by arguing that children who have difficulty with mathematics should be given a restricted diet of basic numeracy skills while allowing only those with a more natural bent for the subject to delve deeper into its secrets and to enjoy the beauty of its patterns. During the middle and late twentieth century many generations of schoolchildren not considered worthy of higher things were given a diet of 'special arithmetic' and expected to feel grateful for it. We are still struggling to come to terms with a postmodern view of what mathematics is and who should be able to benefit from it.

The biologist Lancelot Thomas Hogben, FRS (1895–1975) was a twentieth-century Robert Recorde who had a vision of making the delights of both mathematics and science more widely accessible to the general public. He published *Mathematics for the Million* in 1936, having written the work during a prolonged period in hospital, and this was shortly to be followed by *Science for the Citizen*, published in 1938. Following a distinguished academic career that had included developing the widely used 'Hogben pregnancy test', Lancelot Hogben retired as professor of medical statistics at the University of Birmingham in 1961. He had always maintained a keen interest in popularising the sciences and ensuring that scientific and mathematical ideas were not the exclusive property of a small band of experts but were available for all. He derided those mathematicians who, in his words, were 'inclined to keep the high mysteries of their Pythagorean brotherhood to themselves'. His books promoting an understanding of science and of mathematics enjoyed a wide international readership and ran into numerous editions over many years. Not for him the notion of mathematics for the privileged or mathematics exclusively for high flyers.

The information technology revolution presents new opportunities to help us to rid ourselves of these deeply ingrained prejudices and finally to liberalise mathematics for all citizens. In the spirit of the new methods of communication made available by the new technology I was tempted, like many others, to explore how to use it to open up mathematics to a wider constituency. My chosen vehicle was Twitter.

Is it possible to have meaningful mathematical exchanges on Twitter? That was the challenge that I set myself or, rather, that my grown-up children set me: could their father exploit social media to engage with others in mathematical thinking?

The impetus for this idea came in the wake of the publication in 2012 of the original Welsh version of this book. That book, as does this, includes a smattering of mathematical puzzles that link to each chapter's theme. These proved to be popular and the experiment on Twitter aimed to build on that popularity by posting a puzzle early each weekday morning with the answer posted late that same evening. Each puzzle appears in both English and Welsh, the English versions carrying the hashtags #puzzleoftheday and

#puzzleofthedayanswer. I began the experiment towards the end of August 2012, fully expecting that it would fizzle out after a few weeks. Not so! The juggernaut rolled on, fuelled by the enthusiasm of its followers and a seemingly endless flow of puzzles, original and not so original, each one trying to link in some way with the date on which it is set. For example:

10/01   Which is the best bargain on the shelves, '10% off' or '10% extra free'?

18/01   Which two sets of consecutive numbers add to give a total of 18?

05/03   I have two egg timers – one 5' and one 3'. How can I time a 4' egg?

12/12   The square number 12 × 12 = 144 ends in two 4s. What's the smallest square number that ends in three 4s?

I became intrigued by some of the patterns that appeared to be emerging from this unscientific experiment. For example, I got a strong impression, which is impossible to verify precisely, that the puzzles attracted more female followers than males, many of them being professional women. Those who analyse trends in the use of social media suggest that females represent about 53 per cent of Twitter users. My own analysis of the followers of #puzzleoftheday suggests that about 55 per cent of them are female. Why should that be?

Some solvers appear to be attracted to the puzzles as the equivalent of quick shots of early-morning caffeine to get their brain cells into gear. Most followers choose not to answer on Twitter, preferring to remain anonymous, as one female solver shared with me: 'I like the challenge of trying to do the puzzle without any external pressure; there's no one looking over my shoulder; no one demanding an answer.' Is this the reaction of someone who, for a variety of reasons, failed to realise her potential in mathematics while in school but now gets some satisfaction flexing her mathematical muscles outside the confining walls of the classroom? Can the medium provide a form of escape route from the sterility of sums?

In terms of the National Curriculum the mathematical content of the puzzles is no higher than Level 5, roughly the level attained by the end of primary school or early at secondary school, but solvers need to have a certain curiosity about patterns and to enjoy being challenged. The followers also include schools and teachers, some of whom post their answers in the form of digital images of the work done by their pupils that day, taking the puzzle as their starting point.

It's difficult to draw firm conclusions about the influence of social media – what may be true today may well not stand up to scrutiny tomorrow. Media such as Twitter can open up new opportunities that are accessible to all: they don't discriminate; they don't set entry conditions; and they are non-judgemental. In mathematics, they open up new opportunities to get involved in explorations at a level set by the user. In that sense they can democratise mathematics and remove it – lock, stock and barrel – from the confines of a lofty academia. They have the potential to count us all in.

Largest number puzzle

The number 'three' repeats the letter 'e', 'nine' repeats the letter 'n'. What's the largest number in English that doesn't have any repeating letters?

APPENDIX: THE WELSH NUMBERING SYSTEM

following J. Elwyn Hughes, *Canllawiau Ysgrifennu Cymraeg* (Llandysul, 1998, 2nd edn 2006), pp. 11.4–11.6

CARDINAL NUMBERS 1–100

Number	Traditional/vigesimal	Modern/decimal	English
1	un	un	one
2	dau/dwy	dau/dwy	two
3	tri/tair	tri/tair	three
4	pedwar/pedair	pedwar/pedair	four
5	pump/pum	pump/pum	five
6	chwech/chwe	chwech/chwe	six
7	saith	saith	seven
8	wyth	wyth	eight
9	naw	naw	nine
10	deg/deng	deg/deng	ten
11	un ar ddeg	un deg un	eleven
12	deuddeg/deuddeng	un deg dau/dwy	twelve
13	tri/tair ar ddeg	un deg tri/tair	thirteen
14	pedwar/pedair ar ddeg	un deg pedwar/pedair	fourteen
15	pymtheg/pymtheng	un deg pump	fifteen
16	un ar bymtheg	un deg chwech	sixteen
17	dau/dwy ar bymtheg	un deg saith	seventeen
18	deunaw	un deg wyth	eighteen
19	pedwar/pedair ar bymtheg	un deg naw	nineteen
20	ugain	dau ddeg	twenty
21	un ar hugain	dau ddeg un	twenty-one
22	dau/dwy ar hugain	dau ddeg dau/dwy	twenty-two

Number	Traditional/vigesimal	Modern/decimal	English
23	tri/tair ar hugain	dau ddeg tri/tair	twenty-three
24	pedwar/pedair ar hugain	dau ddeg pedwar/pedair	twenty-four
25	pump ar hugain	dau ddeg pump	twenty-five
26	chwech ar hugain	dau ddeg chwech	twenty-six
27	saith ar hugain	dau ddeg saith	twenty-seven
28	wyth ar hugain	dau ddeg wyth	twenty-eight
29	naw ar hugain	dau ddeg naw	twenty-nine
30	deg ar hugain	tri deg	thirty
31	un ar ddeg ar hugain	tri deg un	thirty-one
32	deuddeg ar hugain	tri deg dau/dwy	thirty-two
33	tri/tair ar ddeg ar hugain	tri deg tri/tair	thirty-three
34	pedwar/pedair ar ddeg ar hugain	tri deg pedwar/pedair	thirty-four
35	pymtheg ar hugain	tri deg pump	thirty-five
36	un ar bymtheg ar hugain	tri deg chwech	thirty-six
37	dau/dwy ar bymtheg ar hugain	tri deg saith	thirty-seven
38	deunaw ar hugain	tri deg wyth	thirty-eight
39	pedwar/pedair ar bymtheg ar hugain	tri deg naw	thirty-nine
40	deugain	pedwar deg	forty
41	un a deugain	pedwar deg un	forty-one
42	dau/dwy a deugain	pedwar deg dau/dwy	forty-two
43	tri/tair a deugain	pedwar deg tri/tair	forty-three
44	pedwar/pedair a deugain	pedwar deg pedwar/pedair	forty-four
45	pump a deugain	pedwar deg pump	forty-five
46	chwech a deugain	pedwar deg chwech	forty-six
47	saith a deugain	pedwar deg saith	forty-seven
48	wyth a deugain	pedwar deg wyth	forty-eight
49	naw a deugain	pedwar deg naw	forty-nine
50	hanner cant	pum deg	fifty
51	hanner cant ac un	pum deg un	fifty-one
52	hanner cant a dau/dwy	pum deg dau/dwy	fifty-two

Number	Traditional/vigesimal	Modern/decimal	English
53	hanner cant a thri/thair	pum deg tri/tair	fifty-three
54	hanner cant a phedwar/phedair	pum deg pedwar/pedair	fifty-four
55	hanner cant a phump	pum deg pump	fifty-five
56	hanner cant a chwech	pum deg chwech	fifty-six
57	hanner cant a saith	pum deg saith	fifty-seven
58	hanner cant ac wyth	pum deg wyth	fifty-eight
59	hanner cant a naw	pum deg naw	fifty-nine
60	trigain	chwe deg	sixty
61	un a thrigain	chwe deg un	sixty-one
62	dau/dwy a thrigain	chwe deg dau/dwy	sixty-two
63	tri/tair a thrigain	chwe deg tri/tair	sixty-three
64	pedwar/pedair a thrigain	chwe deg pedwar/pedair	sixty-four
65	pump a thrigain	chwe deg pump	sixty-five
66	chwech a thrigain	chwe deg chwech	sixty-six
67	saith a thrigain	chwe deg saith	sixty-seven
68	wyth a thrigain	chwe deg wyth	sixty -eight
69	naw a thrigain	chwe deg naw	sixty -nine
70	deg a thrigain	saith deg	seventy
71	un ar ddeg a thrigain	saith deg un	seventy-one
72	deuddeg a thrigain	saith deg dau/dwy	seventy-two
73	tri/tair ar ddeg a thrigain	saith deg tri/tair	seventy-three
74	pedwar/pedair ar ddeg a thrigain	saith deg pedwar/pedair	seventy-four
75	pymtheg a thrigain	saith deg pump	seventy-five
76	un ar bymtheg a thrigain	saith deg chwech	seventy-six
77	dau/dwy ar bymtheg a thrigain	saith deg saith	seventy-seven
78	deunaw a thrigain	saith deg wyth	seventy-eight
79	pedwar/pedair ar bymtheg a thrigain	saith deg naw	seventy-nine
80	pedwar ugain	wyth deg	eighty
81	un a phedwar ugain	wyth deg un	eighty-one
82	dau/dwy a phedwar ugain	wyth deg dau/dwy	eighty-two
83	tri/tair a phedwar ugain	wyth deg tri/tair	eighty-three
84	pedwar/pedair a phedwar ugain	wyth deg pedwar/pedair	eighty-four
85	pump a phedwar ugain	wyth deg pump	eighty-five

86	chwech a phedwar ugain	wyth deg chwech	eighty-six
87	saith a phedwar ugain	wyth deg saith	eighty-seven
88	wyth a phedwar ugain	wyth deg wyth	eighty-eight
89	naw a phedwar ugain	wyth deg naw	eighty-nine
90	deg a phedwar ugain	naw deg	ninety
91	un ar ddeg a phedwar ugain	naw deg un	ninety-one
92	deuddeg a phedwar ugain	naw deg dau/dwy	ninety-two
93	tri/tair ar ddeg a phedwar ugain	naw deg tri/tair	ninety-three
94	pedwar/pedair ar ddeg a phedwar ugain	naw deg pedwar/pedair	ninety-four
95	pymtheg a phedwar ugain	naw deg pump	ninety-five
96	un ar bymtheg a phedwar ugain	naw deg chwech	ninety-six
97	dau/dwy ar bymtheg a phedwar ugain	naw deg saith	ninety-seven
98	deunaw a phedwar ugain	naw deg wyth	ninety-eight
99	pedwar/pedair ar bymtheg a phedwar ugain	naw deg naw	ninety-nine
100	cant	cant	hundred

Notes:

a)   three Welsh numbers have both masculine and feminine forms: 2 *dau/dwy*; 3 *tri/tair*; 4 *pedwar/pedair*.

b)   the first letter of a word in Welsh may change according to what precedes the word. For example: 'four' = *pedwar*, but 'and four' = *a phedwar*; 'ten' = *deg*, but 'two tens' = *dau ddeg*.

ANSWERS TO PUZZLES

CHAPTER 1
Triangle puzzle
There are 13 triangles in the diagram: 1 large, 3 medium and 9 small. Mathematics includes the study of shapes and forms (geometry) as well as the study of numbers (arithmetic).

CHAPTER 2
Explorer's puzzle
The explorer must have started from the North Pole. All journeys in a straight line from the North Pole go due south, a fact that doesn't hold true for any other starting point on the globe, be it in north Canada or in New Zealand or anywhere else. The explorer travels along the three sides of a triangle. The geometry of triangles on a sphere, like the shape of the globe, differs from the geometry of triangles on a flat surface. For example, on a flat surface the sum of the angles of a triangle is 180 degrees. That isn't the case for a triangle on the surface of a sphere. The sum of the angles of the triangle walked by the explorer is greater than 180 degrees.

CHAPTER 3
Roman numerals puzzle
When expressed in Roman numerals, the longest year so far has been 1888, which, as MDCCCLXXXVIII, contains 13 letters.

CHAPTER 4
A weighty puzzle
The cost of carrying the weight is 200 pence (or 16 shillings and 8 pence, remembering that there were 12 pence in a shilling). The question is based on the idea of proportion. The weight of the load having been increased by a factor of five (500/100) from 100 pounds to 500 pounds, the cost of carriage must also increase by the same factor. The length of the journey having been increased from 30 miles to 100 miles the cost must also increase in the same proportion (100/30). We must therefore increase the total cost in two stages, by multiplying 12 pence by 5 and then by 100/30 (or $3^1/_3$). We therefore get $12 \times 5 \times 3^1/_3 = 200$.

CHAPTER 5
Puzzle of the 'Double Rule of Three'
The 'Double Rule of Three' is the label given to the technique used to solve the puzzle in chapter 4. It's a very old expression that first appeared in English in Robert Recorde's book on arithmetic, *The Ground of Artes* (1543). Versions of it can be traced back much further to the early works of mathematicians in the Middle and Far East. The expression was still in use in Britain early in the twentieth century. Does it ring any bells from your own schooldays? That's not very likely and yet here it is being used in a popular song from the late 1960s.

CHAPTER 6
Mayan puzzle
Using Hindu–Arabic numerals this sum represents 11 subtract 3, which is 8. The symbol used by the Mayans to represent this number was:

CHAPTER 7
Vigesimal puzzle
The Welsh word *chweugain* (six twenties) was commonly used before decimalisation in 1971 to refer to ten shillings, because ten

shillings was equivalent to 120 old pence, and so a ten-shilling note was referred to as a *papur chweugain*. Following decimalisation the word *chweugain* continued in use as shorthand for 'fifty pence', and the word is still used today, albeit less frequently, as in the expression *dwy bunt a chweugain* (two pounds and fifty pence).

CHAPTER 8
Prairie puzzle
A horse costs 2,400 pesos more than a sheep. Some algebra can help. If we use H to represent the price of a horse and S the price of a sheep, the question tells us that the difference between (5H + 3S) and (3H + 5S) is 4,800 pesos. But the difference between (5H + 3S) and (3H + 5S) is (2H – 2S), which is double (H – S). It follows that 4,800 pesos is double the difference in price between a horse and a sheep. A price of a horse must therefore be 2,400 pesos more than the price of a sheep. We don't have to know the price of horses or of sheep, simply the difference between them.

CHAPTER 9
Palindrome puzzle
The next palindromic year will be 2112; the first palindromic time after midnight on a digital clock is 01:10.

1, 2, 3 puzzle
Excluding two zeros, the only two whole numbers whose sum is equal to their product are 2 and 2: $2 + 2 = 2 \times 2$.

The only four whole positive numbers whose sum is equal to their product are 1, 1, 2 and 4: $1 + 1 + 2 + 4 = 1 \times 1 \times 2 \times 4$. There are other possibilities if you allow the use of negative numbers (for example: –5, –3, 0, 3, 5) or the use of decimals (for example: 0.5, 1, 2.5, 16).

CHAPTER 10
π puzzle
Pi Day has been celebrated annually on 14 March since 1988. The brainchild of Larry Shaw, a physicist at the San Francisco Exploratorium, the date 14 March was chosen because the American

pattern of writing dates is to put the month before the day, so that 14 March is written as 3/14, corresponding to the pattern of the first three digits of π, 3.14 (three point one four). In 2009, the US House of Representatives passed a resolution recognising 14 March as National Pi Day. The date has attracted increasing worldwide publicity and is celebrated in a vast variety of ways, particularly in schools and colleges, and involves the inevitable consumption of all kinds of pies as well as competitions to memorise and recite as many of the digits of π as possible. Pi Day in 2015 was particularly significant because the date corresponded to the first five digits of π, 3.1415. In recognition of the work of William Jones the Welsh Government took advantage of the special circumstances of 2015 to designate 14 March as Pi Day Cymru, encouraging school pupils and others to develop the celebrations into an annual event.

CHAPTER 11
Binary puzzle
The number that is one less than the binary number 1000 is the binary number 111. In this system the number 1000 corresponds to $(1 \times 2^3) + (0 \times 2^2) + (0 \times 2) + 0$, which is 8. One less than that is 7. We can write 7 as $(1 \times 2^2) + (1 \times 2) + 1$. It follows that 7 is 111 in binary.

CHAPTER 12
Patchwork quilt puzzle
The curate has six sisters. The underlying mathematical theme of this puzzle is common to puzzle books, old and new, but a modern version of it would certainly remove the strong stereotyping that runs throughout the wording of this original version. The solution depends on knowing something about this sequence of numbers: 0, 1, 3, 6, 10, 15, 21, 28, 36, 45, 55, 66, 78,... Note that the difference between consecutive numbers in the sequence increases by 1 each time. The second number (1) is the number of kisses between two girls; the third number (3) is the number of kisses between three girls; the fourth number (6) is the number of kisses between four girls; and so on. The total number of kisses, including those

given by the curate, is 72. The number in the sequence closest to 72 is 66, which is the twelfth number in the sequence. It follows that there must have been 66 kisses between the girls, and that there was a total of twelve girls. That leaves six kisses for the curate, one for each of his six sisters.

CHAPTER 13
Largest number puzzle

The largest number in English that doesn't have any repeating letters is 'five thousand'. The answer to the same question using Welsh number words is *wyth can mil* (eight hundred thousand).

NOTES ON CHAPTERS

I have drawn on a number of sources in writing each chapter, some of which are noted below. I have also added further information and comments for the interested reader.

All websites referred to in these notes were accessed in July 2015.

1. MORE CABBAGE, ANYONE?

A clear distinction between *knowing how* and *understanding why* in the context of learning mathematics was first made by Richard Skemp (1919–95). Skemp highlighted this distinction in 1976 in a very readable article in which he referred to *knowing how* as 'instrumental understanding' and to *understanding why* as 'relational understanding'. Originally published as 'Relational understanding and instrumental understanding', *Mathematics Teaching*, 77 (December 1976), 20–6, this one short article was required reading for a whole generation of mathematics educators and its ideas remain influential. It forms the core of what little educational psychology has informed this book.

A person's attitude to mathematics influences his or her ability to develop an understanding of the subject. The literature on the subject is extensive and persuasive. See, for example, Reuben Hersh and Vera John-Steiner, *Loving + Hating Mathematics* (Princeton, NJ, 2011) and Jo Boaler, *The Elephant in the Classroom: Helping Children Learn and Love Maths* (London, 2009). One of the first to discuss attitudes to mathematics was Laurie Buxton in *Do You Panic About Maths?* (London, 1981). It was this book that first inspired my 'sadistic experiments' with students.

2. MEETING OF MINDS

This chapter is an adaptation into English of an article that was first published in Welsh in the Winter 2010 issue of *Taliesin*, volume 141.

For a further discussion of the interplay between language and mathematically linked concepts see Guy Deutscher, *Through the Language Glass* (London, 2010).

Gender-related perspectives on mathematics have been extensively explored and are discussed further in chapter 12. The quotation in this chapter is from an article by Dorothy Buerk, 'An experience with some able women who avoid mathematics', *For the Learning of Mathematics*, 3/2 (1982), 19–24.

3. 'NOTHING WILL COME OF NOTHING'

The account of Ramanujan's life and work and his relationship with G. H. Hardy is given in David Leavitt, *The Indian Clerk* (London, 2007).

One of the texts that presents traditional Indian mathematics in modern form is that edited by V. S. Agrawala based on Jagadguru Swāmī Śrī Bhāratī Kṛṣṇa Tirthajī Mahārāja, *Vedic Mathematics* (Delhi, 1992). I bought a copy of the book in 2009 at a shop in the foyer of a modern hotel in Bengaluru (previously Bangalore). The shop assistant, a young man, understood nothing of its content.

Angela Saini discusses the duality within modern India in her book, *Geek Nation: How Indian Science is Taking Over the World* (London, 2011).

4. SETTING THE RECORDE STRAIGHT

The association of lecturers in mathematics education holds an annual conference at Gregynog, near Newtown in mid Wales. The 2008 conference celebrated the life and work of Robert Recorde on the 450th anniversary of his death. The University of Wales Press has published a collection of articles based on the conference papers in Gareth Roberts and Fenny Smith (eds), *Robert Recorde: The Life and Times of a Tudor Mathematician* (Cardiff, 2012, 2013). Jack Williams, one of the contributors to the conference, has researched Recorde's life and work extensively. See Jack Williams, *Robert Recorde: Tudor Polymath, Expositor and Practitioner of Computation* (London, 2011). Robert Recorde's life and work is the subject of the first volume in the series of popular books, *Scientists of Wales*, published by the University of Wales Press. See Gordon Roberts, *Robert Recorde: Tudor Scholar and Mathematician* (Cardiff, 2016).

Facsimile copies of Recorde's books have been published by TGR Renascent Books. Details are available on the publisher's website, *www.*

renascentbooks.co.uk. You can also read the original publications online through the websites of major libraries, including the National Library of Wales, and following links to Early English Books Online.

In 2001 a Robert Recorde slate memorial was erected at the entrance to the Robert Recorde Room in the Department of Computer Science at Swansea University. The plaque was designed by John Howes and carved by the calligrapher Ieuan Rees. An image of the plaque can be seen on the Department's website at *www.swansea.ac.uk/compsci/dept/recorde/*.

5. 'NEITHER A BORROWER NOR A LENDER BE'

The phrase 'borrow and pay back' is a meaningless mantra. It is, nevertheless, possible to explain why the method works, something along the following lines. The basis of the 'borrow and pay back' method is that we add ten (10) to both numbers and, in so doing, notice that adding 10 doesn't change the difference between them. The 'trick' is then to add 10 to the top number in the units column and to add 10 to the bottom number in the tens column. The technical term for this operation is 'equal addition' because we are adding 10 equally to both numbers. This explanation isn't easy, even for an adult. Imagine trying to explain it to a small child. It's therefore not completely surprising that many teachers felt uncertain and that the terms 'borrow' and 'pay back' came to be adopted in order to help children to *know how* while ignoring the need for them also to *understand why*. An argument that was often used to justify the continued use of the 'borrow and pay back' method was that it was neater than the 'decomposition' method when using 'logarithm tables', a method of calculation that was very common before calculators became popular from the 1970s onwards. By today, log tables have long since disappeared from classrooms.

John Thomas wasn't alone in publishing books in Welsh on mathematics during the Industrial Revolution. In the chronological order of their first editions, the main publications were: John Roberts, *Arithmetic: mewn Trefn Hawdd ac Eglur*, 1768; John Thomas, *Annerch i Ieuengctyd Cymru*, 1795; Thomas Jones, *Rhifiadur*, 1827; John William Thomas (Arfonwyson), *Elfennau Rhifyddiaeth*, 1831. In his book *Mathemategwyr Cymru* (Cardiff, 1994), Llewelyn Gwyn Chambers provides a comprehensive overview of the history of Welsh mathematicians, including writers of mathematical texts in Welsh.

The position adopted by many politicians in relation to mathematics often reveals their wider views of education in general. A classic example is their take on 'long division'. When the National Curriculum was first introduced in England and Wales in 1988, mathematics educators were keen to avoid including the traditional methods of doing long-division calculations. The notion of division is essential, of course, and children have to develop a range of methods for performing calculations that involve division, but educators at that time argued that children were unable to understand the traditional way of doing long division and that, as a result, they took a dislike to mathematics more generally. The professional opinion was virtually unanimous. However, Margaret Thatcher, then prime minister, disagreed. For her, keeping the traditional method was equated with maintaining 'standards' and any thought of being rid of it would undermine all that Britain stood for. It was a bizarre intervention by a politician, let alone a prime minister, but Margaret Thatcher remained adamant, and a compromise position was eventually agreed whereby the sacred cow of long division was kept in the curriculum but schools were allowed to introduce it in a variety of different ways. That arrangement persists to this day and the occasional London-based education minister, particularly on the Tory benches, repeats the mantra from time to time equating long division with some kind of 'gold standard'.

In some of their popular 1940s films, the American comedians Abbott and Costello satirised the mechanical rote methods of doing sums. One of their most popular sketches, available intermittently on the internet, describes how half-remembered gobbledygook can be used to show indisputably that thirteen sevens is twenty-eight ($13 \times 7 = 28$).

6. AMAZING MAYANS

Can we explain why words such as *deuddeg* (twelve), *pymtheg* (fifteen) and *deunaw* (eighteen) entered Welsh vocabulary and have stood the test of time, even though they were originally linked with the vigesimal system but have since developed a life of their own and continue to be used in modern Welsh alongside the more usual decimal words? It is possible that they originated in parallel with the development of methods of counting money and measuring in agriculture and trade and may also have been influenced by the development of oral methods of keeping a tally. For example, Dafydd Wyn Jones remembers his father, Simon Jones, a farmer from Llanuwchllyn near Bala early in the twentieth century, counting sheep in threes as far as thirty: *tair, chwech, naw, deuddeg, pymtheg, deunaw, un ar hugain, pedair ar hugain, saith ar hugain, deg ar hugain* (3, 6, 9, 12, 15, 18, 21, 24, 27, 30). The words flow easily, particularly the *deuddeg, pymtheg, deunaw*. Things get more complicated after 30 and Simon Jones would start again with *tair, chwech, naw* etc. at that point, keeping count of how many 30s were in his total. Shepherds in the north of England are also known to have used similar words for counting, including *bumfit* corresponding to the Welsh *pymtheg* for fifteen. It was also common practice to count loaves in threes in the bakehouse (the span of a hand enabling the baker to touch three loaves at a time), and to count eggs in threes (because a hand could hold three eggs at a time). Both loaves and eggs would then be counted in dozens, counting: *tri, chwech, naw, dwsin; tri, chwech, naw, dau ddwsin,...* (three, six, nine, dozen; three, six, nine, two dozen,...). When dealing with money the term *deunaw* (eighteen) was a neat label for one shilling and sixpence, corresponding to 18 pence. Dafydd Wyn Jones recalls his father telling a story of two farmers disagreeing over the number of sheep in a field, one claiming that there were *tair ar bymtheg* (three on fifteen) and the other that there were *deunaw* (two nines).

Examples of the oral use of the vigesimal system have been collated by Dr Ceinwen Thomas, an authority on Welsh dialects. Ceinwen Thomas was born and bred in Nantgarw, a village in the Taff Valley, halfway between Cardiff and Pontypridd, and took particular delight in the dialects associated with the valleys of south-east Wales, particularly her own Gwernhwyseg. Another expert in linguistics who recorded examples of the use of the vigesimal system was O. H. Fynes-Clinton, Professor of French at the University College of North Wales, Bangor. One of his fields of research was the dialects used in and around Bangor, his work published as *The Welsh Vocabulary of the Bangor District* (London, 1913).

Counting in German provides an unexpected example of a modern controversy regarding the use of the German equivalent of twenty – *zwanzig*. In German the traditional way of expressing the number 21 is *einundzwanzig*, equivalent to 'one and twenty' and similar to the traditional Welsh *un ar hugain*, a method of counting that gives the number of units before the number of tens. For example the German for 47 is *siebenundvierzig* – seven and forty. In 2004 a group of German academics at the University of Ruhr Bochum began a campaign to introduce a new way of saying numbers, not necessarily to displace the original method, but at least to have equal parity with it. In the new method 21 would be *zwanzigeins* (twenty one) and 47 would be *vierzigsieben* (forty seven). The academics argued their case on the basis of the 'educational and economic' advantages that would flow from a system that changes the traditional way of saying numbers to a way that more closely corresponds to the way in which numbers are written. See *www.verein-zwanzigeins.de*.

A lively canter through the idiosyncrasies of European languages, including their words for numbers, is provided by Gaston Dorren, *Lingo: A Language Spotter's Guide to Europe* (London, 2014).

7. WHAT DO YOU RECKON?

The Welsh-language weekly paper *Seren Gomer* began publication in 1814 and proved to be an instant success, covering international and national news as well as information about local fairs and markets. It also featured a popular column of readers' letters, many of their authors adopting pseudonyms. In 1820 a spate of letters appeared debating the pros and cons of Welsh counting methods, the vast majority of them in favour of the adoption of a new decimal method. The story at the beginning of this chapter is based on a letter from one styling himself as '*Aigiochus o Fyllin*' that refers to 'a counting competition, the result of which was in doubt over the first hundred numbers, but by the time two hundred was reached, the sons of Gomer jubilantly roared out their VICTORY, which echoed around the streets and caused much surprise to the English'. (my translation)

When the Reverend David Jones printed a second edition of his hymnal in 1821, the number of hymns in the volume now reaching as many as '*saith gant a thrigain*' ('seven hundred and three twenties', 760), he took the bold decision to print Hindu–Arabic numerals, such as 492, above each one. By the time that he published his third edition in 1827 he had developed cold feet, and was persuaded by the weight of tradition to revert to Roman numerals.

During the 1970s there was much heated debate in the Welsh-language press regarding the changes in the language of counting that were by then being widely used in schools. Some, such as Dr Ceinwen Thomas (see above), were vociferous in their support for using the traditional vigesimal system in the classroom. They argued that the traditional way of counting was deeply rooted in the Welsh oral tradition and that, by contrast, the newfangled decimal numbers merely aped English numbers. Although Iorwerth Peate also supported keeping to the traditional methods, he sought a compromise solution whereby children could be taught their sums using the new system in the classroom but that the traditional system should reign supreme outside that context. Such a compromise would be unnatural and impractical, but the fact that it was raised in good faith as a possibility by Iorwerth Peate was, in itself, a sign that change was inevitable.

There has been some confusion regarding how to tell the time in Welsh. In his book *Modern Welsh: A Comprehensive Grammar* (London, 2003), Gareth King advises learners of Welsh as follows: 'It is important to think of time in Welsh as a clock face rather than numbers. We cannot say tri pumdeg pump for 3.55, as we can in English.' Even if such a statement could be upheld in 2003 the argument does not ring true today. The grammarian J. Elwyn Hughes points out that 'when reading an analogue clock the traditional method of telling the time is used ... but that, when reading a digital clock, both methods can be used and there is a tendency to turn to the modern method when things get complicated!' (*Canllawiau Ysgrifennu Cymraeg* [Llandysul, 2006], p. 11.1). (my translation)

In 1998 the Welsh Language Board, then the statutory body that promoted the use of Welsh, published guidelines regarding how to write numbers in Welsh on bank cheques. A second draft of the guidelines was published to coincide with the new millennium and 10,000 copies were distributed to staff and customers in banks and retail outlets.

In 1999, Aled Glynne Davies, then editor of the Welsh-language service BBC Radio Cymru, produced guidelines for broadcasters. In the section on numbers he suggested the following compromise:

> Up to the number 30, it is acceptable to say *un ar hugain* (one on twenty), *dau ar hugain* (two on twenty) and so on. But after *deg ar hugain* (ten on twenty), you should say '*tri deg un, tri deg dau*' (three tens one, three tens two), and so on ... The number 31 presents a particular difficulty when giving the date. If, for example, you are referring to 31 January, you should say '*diwrnod olaf Ionawr*' (the last day of January). The form '*Ionawr tri deg un*' (January three tens one) is also acceptable. (my translation)

> Note: saying 31 January in Welsh using the traditional vigesimal sys-
> tem is no mean feat as it takes the form '*Ionawr yr unfed ar ddeg ar
> hugain*' ('January the first on ten on twenty').

The results of research that demonstrate the advantages of learning
to count in Welsh are published in an article by Ann Dowker and Delyth
Lloyd, 'Linguistic influences on numeracy' (2005). The article is
available online on the website of Bangor University's School of Education:
www.bangor.ac.uk/addysg/publications/Mathematics_Primary_School.pdf.
Malcolm Gladwell's *Outliers* (London, 2008), pp. 227–32, includes a discussion of learning to count in the languages of the countries of the Far East.

8. PRAIRIE POWER

The first group of Welsh settlers set sail in May 1865 on board the *Mimosa*
from Liverpool, landing two months later in what appeared to be an
inhospitable desert some 850 miles south of Buenos Aires. They established a community in what is now referred to as *Dyffryn Camwy* (the
valley of the river Camwy) and extended this community some years
later to include a settlement in *Cwm Hyfryd* (literally, the beautiful valley) in the Andes, 400 miles west of their original base, encompassing
the towns of Esquel and Trevelin. Both settlements are in Patagonia, a
large and sparsely populated region that spans the southern parts of both
Chile and Argentina, south of the Rio Negro. The settlers are referred to as
Patagonians in this chapter, there being no accepted English equivalent
of the Welsh collective term *Gwladfawyr*. The descendants of those first
settlers and others from Wales who followed over the years have maintained their Welshness while also embracing an Argentine identity. They
are Argentinians, first and foremost, whose first language is Spanish, but a
significant proportion of them have kept the Welsh of their forefathers or
have learned Welsh as a second language. It is remarkable that, 150 years
after the first landing at what is now called Puerto Madryn, the Welsh
language remains an important part of the Patagonian cultural mix. Most
of the available published material about the history of the Welsh settlement is in Welsh and/or Spanish. A trilingual website (including English)
has been developed jointly by the National Library of Wales and Bangor
University's Library and Archives Service, with the support of the Welsh
Government: *www.glaniad.com*. The Wales-Argentina Society was founded

in 1939 by a number of people with connections in the Welsh settlement in Chubut to be a link between the two countries. The society maintains a trilingual website: *www.cymru-ariannin.org.*

9.   PUTTING DOWN DIGITAL ROOTS

You can spend many hours trawling the web for examples of palindromes. The website *www.palindromelist.net* lists some hundreds of English palindromic words and palindromic sentences. An interesting example was created by Peter Hilton, a member of the Bletchley Park wartime code-cracking teams who was subsequently appointed Mason Professor of Pure Mathematics at the University of Birmingham before going on to hold a number of chairs in America. His 51-letter palindrome is: 'Doc note, I dissent. A fast never prevents a fatness. I diet on cod.' In music, *The Palindrome* is the informal title of Joseph Haydn's 47th Symphony, the third movement of the symphony being a musical palindrome.

Jen Jones and others have led a revival in interest in Welsh quilts. Jen Jones runs a quilt shop in Llanybydder, Ceredigion, and a permanent exhibition of quilts in the old Town Hall in nearby Lampeter. Details of both the shop and the exhibition are on the website *www.jen-jones.com.* The Quilt Association is based in the Minerva Arts Centre, Llanidloes, Powys, and maintains the website *www.quilt.org.uk.* See also Jen Jones, *Welsh Quilts* (Carmarthen, 1997).

It is one thing to spot what may be a number pattern; it is quite another thing to prove indisputably that the pattern holds universally. In this chapter the pattern in the 9-times table led us to surmise that 'the digital root of *every number* in the 9-times table is 9'. We experimented with some specific examples – the digital root of 36 is 9, the digital root of 144 is 9, and so on – and each example confirmed our theory. But how can we be absolutely sure that the statement holds true for every number in the 9-times table, and not just for those specific examples? And how can we establish that truth given that we are clearly unable to list every possible number in the table? In order to provide a cast-iron proof we would need to use some algebra, which would be beyond the scope of this book. However, that does not prevent us from investigating the pattern further, doing more examples, and building up our confidence that our hunch may well be true. Mathematicians rely heavily on this instinct for rooting out what may be a pattern, and only then do they go about trying to find a proof that backs up their hunch. Should you wish to experiment further, a good place to start could be to explore the patterns in the 8-times table. These are the numbers in that table, as far as 10 × 8:

$$8, 16, 24, 32, 40, 48, 56, 64, 72, 80$$

The digital root of 16 is 7 (using the notation used in the chapter, $16 \longrightarrow 7$). Working along the numbers do you agree that this is the sequence of the digital roots of each number in turn?

$$8, 7, 6, 5, 4, 3, 2, 1, 9, 8$$

Can you see a pattern here? It's certainly different to the pattern in the 9-times table. Does the pattern continue as you extend the table further than 80? What about other tables? And you're well on your way as an explorer of number patterns.

Prime numbers provide the basis for the security of our bank cards. The number 13 is prime because we can't divide it by any number less than 13 (apart from 1, of course). The number 14 isn't prime because we can divide it by other numbers, specifically by 2 and by 7. These are the first few prime numbers: 2, 3, 5, 7, 11, 13, 17, 19, 23,... (for reasons of mathematical convenience, 1 is not regarded as being a prime number). The security of bank cards depends on using a code based on very, very large prime numbers – ones that contain tens of digits. Prime numbers have fascinated mathematicians through the ages. See, for example, John Derbyshire, *Prime Obsession* (Washington, 2003) and Marcus du Sautoy, *The Music of the Primes* (London, 2003).

One of the most influential books on mathematics published in the twentieth century was *A Mathematician's Apology*, in which the author, G. H. Hardy, sets out to explain how mathematicians tend to think. The closing sentence of the chapter is an adaptation of Hardy's often-quoted sentence: 'A mathematician, like a painter or poet, is a maker of patterns.' The book was first published in 1940, and a second edition published in 1967. The revised edition includes a preface by C. P. Snow, physicist and novelist, renowned for his 1959 public lecture, 'The Two Cultures'. In that lecture C. P. Snow expresses his concern regarding the gap that he perceived between scientists, on the one hand, and, on the other, scholars specialising in the humanities. Snow's message that the gap needs to be bridged has informed writers ever since and has also influenced the theme of this book.

In Welsh-speaking circles many writers have addressed the issues identified by C. P. Snow, most notably the poet and scholar Sir T. H. Parry-Williams (1887–1975), some of whose essays bridge both sides of the cultural divide. See T. H. Parry-Williams, *Casgliad o Ysgrifau* (Llandysul, 1984).

10. AREAS OF (MIS)UNDERSTANDING

The most common response to the question as to what is the ratio of a circle's circumference to its diameter is the fraction 22/7. The value of this fraction, slightly greater than 3, is indeed close to the true value but it is only an approximation to it. A better approximation is the fraction 355/113, first discovered in China about AD 500.

The value of π correct to 20 decimal places is 3.14159265358979323846. By today mathematicians have been able to calculate the number correct to millions of decimal places. A feat that attracts popular attention from time to time is the challenge to commit as many as possible of these decimal places to memory. *The Guinness Book of Records* reports that the current world champion is Chao Lu, a student from China, who succeeded in remembering the first 67,890 digits of π, reciting them correctly over a period of twenty-four hours on 20 November 2005. The British record is held by Daniel Tammet who, on Pi Day in 2004, recited the first 25,514 decimal places in a little over five hours to an audience at the Museum for the History of Science in Oxford, an event that is memorably described by the record holder in his book, *Thinking in Numbers*, pp. 114–25. There is no shortage of mnemonics to help those with shorter memories, including the snappy 'How I wish I could calculate pi' corresponding to 3.141592 and the slightly longer 'How I like a drink, alcoholic of course, after the heavy lectures involving quantum mechanics', which corresponds to the first fifteen digits of π. Sources of π-related mnemonics on the web include the Wikipedia entry for 'Piphilology' and Mike Keith's extraordinary website *www.cadaeic.net*.

Some of the details of William Jones's life remain unclear, partly because many of his personal papers have not come to light. However, interest in his life and work has increased. For example, Patricia Rothman at University College London has analysed William Jones's circle of influence in London in her article 'William Jones and his circle: the man who invented the concept of pi', *History Today*, 59/7 (2009), 24–30.

The Library and Archives Service at Bangor University holds a number of documents that relate to William Jones. The collection has been augmented over the years by material contributed by Llewelyn Gwyn Chambers (1924–2014), previously a Reader in the university's Department of Mathematics and an ardent promoter of William Jones and his work.

For further details on the life and work of Sir William Jones, the son, see Michael J. Franklin, *Orientalist Jones: Sir William Jones, Poet, Lawyer, and Linguist, 1746–1794* (Oxford, 2011).

11. CRACKING THE CODE

Understanding 'place value' is the basis of work with numbers. Early counting experiences with simple objects like buttons, for example, can be particularly helpful: arrange the buttons in piles of ten; notice that there are 2 piles and another 4 buttons left over; the penny drops when you realise that there are 24 buttons altogether without having to count them 1, 2, 3,... The question about the milometer is far from easy. You need to understand place value and to be familiar with how a milometer works. The correct answer is:

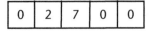

Common wrong answers include:

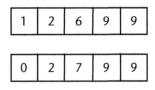

Can you follow the reasoning of a child who gives one or the other of these answers?

The easiest way to try to understand binary numbers is to compare them with the more familiar decimal numbers. Counting in tens, the number 347 means 300 + 40 + 7, which can be written as $(3 \times 10^2) + (4 \times 10) + 7$. Similarly, the number 111 in binary means $(1 \times 2^2) + (1 \times 2) + 1$, a total of 7. Extending this idea further, the binary 10111 is $(1 \times 2^4) + (0 \times 2^3) + (1 \times 2^2) + (1 \times 2) + 1$, which is a total of 23. I had asked Gareth what, in binary, is the number one less than 1000000. The value of the '1' in the binary 1000000 is 2^6, or $2 \times 2 \times 2 \times 2 \times 2 \times 2$, which is 64. One less than that is 63, which can be represented as the binary number 111111. How can we check that? Well, 111111 is $(1 \times 2^5) + (1 \times 2^4) + (1 \times 2^3) + (1 \times 2^2) + (1 \times 2) + 1$, which is $32 + 16 + 8 + 4 + 2 + 1$, a total of 63. You may find that confusing, but it should still give you a feel for Gareth's insight, as a five-year-old, into how the whole system worked. Gareth was very special.

12. DOES MATHEMATICS HAVE A GENDER?

The archive maintained by the University of St Andrews in Scotland containing biographical details of mathematicians from all parts of the world can be viewed online at *www.history.mcs.st-and.ac.uk/index.html* or by searching for 'The MacTutor History of Mathematics archive'.

Mathematics is the theme of issue 141 of the journal *Taliesin* (Winter 2010). Hafina Clwyd's response appeared in her column in the *Western Mail* on 21 December 2010. In her autobiographical books, Hafina Clwyd refers to her mathematics experiences in *Merch Morfydd* (Caernarfon, 1987) and in *Buwch ar y Lein* (Aberystwyth, 1987).

In her book *Women in Mathematics* (Cambridge, MA, 1974), Lynn Osen analyses the impact women have had on the development of mathematical thought. Her premise is set out in the preface to the book:

> Many women in our present culture value mathematical ignorance as if it were a social grace, and they perceive mathematics as a series of meaningless technical procedures. They discount the role mathematics has played in determining the direction of philosophic thought, and they ignore its powerful satisfactions and its aesthetic values, which are equal to those offered by any other branch of knowledge. Yet today more than at any other point in time an appreciation of mathematics is necessary a priori if we are to solve our problems and share in building a better world.

Attitudes have shifted since 1974 but the influence of previous generations remains significant even now.

For further information about the artist Claudia Williams, see Robert Meyricke, *Claudia Williams* (Aberystwyth, 2000), and Harry Heuser and Robert Meyricke, *Claudia Williams: An Intimate Acquaintance* (Bristol, 2013).

13.   HOW TO MAKE MATHS REAL FOR ALL OF US

This chapter is an expanded version of an article that first appeared in *Mathematics in School*, a journal published by the Mathematical Association: 'Twitter maths', 42/4 (September 2013), 37.

Lancelot Hogben enjoyed a rich and productive career. See Robert Bud, 'Hogben, Lancelot Thomas (1895–1975)', *Oxford Dictionary of National Biography* (Oxford, 2004). Bud encapsulates Hogben's abilities and lack of conformity by noting that he 'displayed a brilliance whose rewards were undermined with "a sheer genius for making enemies"'. On retirement from his academic post, Lancelot Hogben moved to live in Glyn Ceiriog, a village in north-east Wales. Having already developed a keen interest in comparative linguistics, at Glyn Ceiriog he quickly learned to read and speak Welsh. He retained a strong sense of the extraordinary, if not the eccentric, and set himself unusual targets. In 1967 he published *Whales for the Welsh*, a novel that has the peculiarity that all the words in the book are monosyllables as, indeed, is the title of this chapter. 'As I got into my stride', recalled Hogben, 'I found that the effort to write without recourse to polysyllables generated a style of its own, and one with an almost inescapably ironical flavour.' An autobiography, edited by members of Hogben's family, was published in 1998, and his papers are held in the archives of the University of Birmingham. Hogben's academic achievements, together with his close associations with Wales, were recognised by the University of Wales in 1963 when he was awarded an honorary DSc.

The author has published three collections in Welsh of samples of the puzzles that have been posted on Twitter: *Posau Pum Munud* (Llandysul, 2013), *Posau Pum Munud 2* (Llandysul, 2014) and *Posau Pum Munud 3* (Llandysul, 2016).

Figures provided by the General Teaching Council for Wales (GTCW) show that approximately three-quarters of all teachers in Wales, 84 per cent of all primary-school teachers and 65 per cent of all secondary-school teachers are female. In science and mathematics, 57 per cent of all secondary-school teachers of mathematics in Wales are female, 37 per cent of teachers of physics, 59 per cent of teachers of chemistry and 65 per cent of teachers of biology. By way of contrast, 84 per cent of all teachers of Welsh at secondary school are female, 83 per cent of teachers of English and 87 per cent of teachers of modern foreign languages. (Data taken from the Register of Qualified Teachers on 1 December 2014.)

FURTHER READING

IN PREPARING this book I have benefited greatly from recent publications that discuss mathematical ideas adopting a populist style. Chief among them are the two books by Alex Bellos: *Alex's Adventures in Numberland* (London, 2010) and *Alex through the Looking-Glass* (London, 2014).

I have also greatly enjoyed the following books, among many others:

Du Sautoy, Marcus, *Finding Moonshine: A Mathematician's Journey through Symmetry* (London, 2008)

Gessen, Masha, *Perfect Rigour: A Genius and the Mathematical Breakthrough of the Century* (London, 2011)

Hoffman, Paul, *The Man who Loved only Numbers: The Story of Paul Erdös and the Search for Mathematical Truth* (London, 1998)

Masters, Alexander, *The Genius in my Basement: The Biography of a Happy Man* (London, 2011)

Singh, Simon, *The Simpsons and their Mathematical Secrets* (London, 2013)

Stewart, Ian, *Professor Stewart's Casebook of Mathematical Mysteries* (London, 2014)

Tammet, Daniel, *Thinking in Numbers* (London, 2012)

Wilson, Robin, *Lewis Carroll in Numberland* (London, 2008)

INDEX

algebra 2, 11, 28–9, 72, 75, 103–5, 122, 133
Archimedes 79–82, 101
arithmetic 5, 19, 27–9, 38, 105, 112, 120–1

Berwyn, R. J. 61–2, 64–6
borrow and pay back *see* sums

Celtic languages 42, 44–5
Chambers, Llewelyn Gwyn 127, 136
circle(s) 80–1, 83, 86–7, 135
Clwyd, Hafina 107–8, 137

Davies, Sydney 102, 104–5
decomposition *see* sums
'Double Rule of Three' *see* sums

Einstein, Albert 16–17
equal addition *see* sums
Euclid 21, 27, 40, 112

geometry 9, 21, 27–8, 40, 64, 105, 109, 112, 120

Hardy, G. H. 16, 68, 126, 134
Hogben, Lancelot Thomas 112–13, 138

Jones, William 80–6, 88, 123, 135–6
his life 82–6, 135–6
use of the symbol π (pi) 80–1, 83

knowing how 4–5, 17, 19, 36, 38, 40, 82, 87–8, 125

Lloyd, O. M. 100

mathematics
authority 2
hatred of 6, 100
and language 5
like cabbage 6
and manhood 97, 99, 106
panic 1, 4
patterns 67–77
and women 97–110
Mayans 41–5, 121

namyn 93–5
Newton, Isaac 16, 84–5
numbers
binary numbers 92–3, 136
decimal system (modern) 18, 41–2, 49, 51–7, 60–4, 94–5, 116, 128, 130–1

deunaw 41–2, 51, 54, 63–4, 95, 128–9
Hindu-Arabic numerals 12, 19, 121, 130
place value 90–2, 136
Roman numerals 18–19, 49, 95, 120, 130
vigesimal system (traditional) 41–5, 49, 51–2, 54, 56–7, 60, 62, 94, 116, 121, 128–30, 132
zero 18, 42, 44, 92, 122

palindrome(s) 67–70, 122, 133
Patagonia 59–66, 132–3
Museo Histórico Regional 64–5
Peate, Iorwerth C. 52–3, 131
Phillips, Morien 74–6
pi (π) 80–3, 85–6, 88, 122–3, 135–6

Ramanujan, Srinivasa 15–17, 126
Recorde, Robert 11–12, 19, 21–32, 38–40, 70, 109, 110, 112, 121, 126–7
Earl of Pembroke 24–5, 30
sign of equality 12, 21, 29–30
Tenby 11, 21–5, 28, 109
The Castle of Knowledge 26
The Ground of Artes 19, 27, 32, 38, 121

The Pathway to Knowledg 26–7
The Whetstone of Witte 11, 27, 30

sums
addition 5, 29
'Double Rule of Three' 40, 121
long division 4–5, 128
multiplication tables 61
subtraction 5, 29, 33–40, 45
borrow and pay back 36–9, 127
decomposition 35–7
equal addition 127

Thomas, Ceinwen 129–30
Thomas, John 39–40, 127

understanding why 5, 17, 28–9, 36, 38, 40, 69, 82, 87–8, 90, 92–3, 125
University of St Andrews 101, 105, 137

Warner, Mary Wynne 101–5
Williams, Claudia 109, 137
Williams, Gareth Wyn 89–95, 136–7

IN MEMORY OF GARETH WYN WILLIAMS
(1978–2010)
MATHEMATICIAN

Gareth Wyn Williams, a son of Anglesey like William Jones, was an outstanding mathematician. In his short career he applied his mathematical skills to combat terrorism as a member of staff at the Government Communications Headquarters (GCHQ) in Cheltenham. During a secondment to MI6 in 2010, at the tender age of 31, Gareth died in tragic and unexplained circumstances.